Mathematics for Technicians 1

Frank Tabberer

Lecturer in Mathematics, Department of Building,
Chichester College of Technology

NEWNES–BUTTERWORTHS
TEC
TECHNICIAN SERIES

THE BUTTERWORTH GROUP

UNITED KINGDOM Butterworth & Co. (Publishers) Ltd
London: 88 Kingsway, WC2B 6AB

AUSTRALIA Butterworths Pty Ltd
Sydney: 586 Pacific Highway, Chatswood, NSW 2067
Also at Melbourne, Brisbane, Adelaide and Perth

CANADA Butterworth & Co. (Canada) Ltd
Toronto: 2265 Midland Avenue, Scarborough, Ontario M1P 4S1

NEW ZEALAND Butterworths of New Zealand Ltd
Wellington: T & W Young Building, 77–85 Customhouse Quay, 1, CPO Box 472

SOUTH AFRICA Butterworth & Co. (South Africa) (Pty) Ltd
Durban: 152–154 Gale Street

USA Butterworth (Publishers) Inc
Boston: 19 Cummings Park, Woburn, Mass. 01801

First published 1978

British Library Cataloguing in Publication Data

Tabberer, Francis Leonard
Mathematics for technicians
1.
1. Shop mathematics
I. Title
510'.2'46 TJ1165 77-30578
ISBN 0-408-00326-X

Typeset by Reproduction Drawings Ltd, Sutton, Surrey
Printed in England by Page Bros Ltd., Norwich, Norfolk

Mathematics for Technicians 1

Preface

This book has been prepared for the use of trainee technicians studying mathematics at Level 1 in courses promoted by the Technician Education Council, and covers fully the work required to achieve the learning objectives contained in TEC unit U75/005, *Mathematics 1*. The presentation has been designed to create an interest in the subject, particularly for those students who may have found mathematics a stumbling block during earlier studies and who may be entering a TEC course as school leavers with limited mathematical attainment. The older student returning to college after a pause in his formal education usually finds difficulty in recalling more than basic everyday arithmetic, and should find this volume a useful stepping stone to the mathematics in the second and third levels of the course.

Each chapter commences with a list of its contents and ends with a summary of the subject matter covered. Frequent exercises are provided throughout the text, and a revision exercise is included at the end of each chapter.

F.T.

Contents

A Manipulating numbers

This chapter deals with:
Integers, directed numbers, rules of signs
Prime numbers, factors, HCF and LCM
Commutative, associative and distributive laws
Vulgar fractions, proper and improper fractions, mixed numbers
Ratio and proportion
Decimal fractions, non-terminating decimals, recurring decimals
Percentages
Powers, roots and reciprocals; indices
Standard form, binary numbers

1 BASIC ARITHMETIC

1.1 Positive and negative integers; factors

Although Roman numerals are still used on date inscriptions and clock faces, it is the Arabic system of numerals that we use to write natural numbers, 1, 2, 3, 4, 5 . . . etc. Each single numeral is a *digit*. The digits from 1 to 9, with a nought, allow the writing of numbers of any magnitude in columns of tens, hundreds, thousands and so on. This is the *denary system* of numbers.

A whole number is an *integer* and may be *positive* or *negative*. If some arbitrary value is chosen as a starting point or *zero*, then greater values are regarded as positive and lesser values as negative. A minus sign (–) indicates negative values. For example, –25°C indicates a temperature below the chosen zero value. A positive value such as 25°C could be written as +25°C, but since positive values are regarded as normal the + sign is omitted. Such numbers, relating to some chosen zero, are *directed numbers.*

The signs + and – are also used to indicate *addition* and *subtraction.*

$5 + 3$ means add 3 to 5 (= 8)
$5 - 3$ means subtract 3 from 5 (= 2)
$-5 - 3$ means subtract 3 from –5 (= –8)
$-5 + 3$ means add 3 to –5 (= –2)

A double negative indicates a reversal of the negative, which becomes positive.

e.g. $5 - (-3) = 5 + 3 = 8$

The rules of signs for addition and subtraction may be summarised as:

+ (+) means +
– (–) means +
+ (–) means –
– (+) means –

i.e. *like signs combine to a positive result, unlike signs combine to a negative result.*

The result of addition is the *sum* of the quantities added, the result of subtraction is their *difference.*

Multiplication is the result of taking a given number a certain number of times. The number taken is the *multiplicand*, and the number of times it is taken is the *multiplier*. The result is the *product*. The rules of signs for multiplication may be derived from the rules for addition. Taking 9 three times may be written as 3×9 or as $9 + 9 + 9$.

$$3 \times 9 = 9 + 9 + 9 = 27$$
$$3 \times -9 = (-9) + (-9) + (-9) = -9 - 9 - 9 = -27$$
$$-3 \times 9 = -(+9) - (+9) - (+9) = -9 - 9 - 9 = -27$$
$$-3 \times -9 = -(-9) - (-9) - (-9) = +9 + 9 + 9 = 27$$

Summarising these results we have:

(+) × (+) gives +
(+) × (–) gives –
(–) × (+) gives –
(–) × (–) gives +

Division is the result of separating a given number into a number of equal parts. The number divided is the *dividend*, and the number of parts into which it is divided is the *divisor*. The result (the value of each part) is the *quotient*.

The rules of signs for division are the same as for multiplication since, as we shall see later, division may be regarded as fractional multiplication.

i.e.
$$8 \div 2 = 4$$
$$8 \div (-2) = -4$$
$$-8 \div 2 = -4$$
$$-8 \div (-2) = 4$$

So, in both multiplication and division, *like signs combine to a positive result and unlike signs combine to a negative result.*

FACTORS

Factors of a number are those smaller numbers that divide into it exactly.

e.g. 3 and 7 are a pair of factors of 21

A *prime number* is one that has no factors except itself and 1.

e.g. 1, 3, 5, 7, 11, 13, 17 . . . etc.

Prime factors are obtained when a number is factorised and then its factors factorised until only prime numbers result.

e.g. $42 = 2 \times 21$ (these are factors of 42)
$= 2 \times 3 \times 7$ (these are the prime factors of 42)

Common factors are those that occur as factors of all the numbers in a group.

e.g. 3 is a common factor of 12, 18, 24, 30 and 36

Such a group may have several common factors, and 2 and 6 are also common factors of the above group.

Highest common factor (HCF) is the greatest number that is a common factor of a group. In the example quoted above, 6 is the HCF of 12, 18, 24, 30 and 36.

The *lowest common multiple* (LCM) of a group of numbers is the least number into which each of the group will divide exactly. Consider the group 4, 6, 8 and 12. Common multiples include 48, 72 and 96 since each of the group will divide into them without a remainder. The LCM however is 24, the *lowest* number into which they will all divide.

Example Express 18, 24 and 36 in terms of their prime factors, and find their HCF and LCM.

$18 = 2 \times 3 \times 3$
$24 = 2 \times 2 \times 2 \times 3$
$36 = 2 \times 2 \times 3 \times 3$

For the HCF use *each prime factor* that is common to all sets *once only* and obtain their product.

$HCF = 2 \times 3 = \underline{6}$

For the LCM use *each prime factor* the *greatest number of times* it appears in any one set.

$LCM = 2 \times 2 \times 2 \times 3 \times 3 = \underline{72}$

Exercise 1a Write down the values of the following:

1. $3 - (+2)$ 2. $-(-3)$ 3. $(-1) - (-1)$
4. $(-7) + (-5)$ 5. $4 + (-6)$ 6. $2 - (+0)$
7. $(-3) \times (+2)$ 8. $4 \times (+6)$ 9. $9 \times (-5)$
10. $(-6) \div (-3)$ 11. $63 \div (-7)$ 12. $(-7) \div 14$

Express the following numbers in terms of their prime factors:

13. 210 14. 231 15. 1 690 16. 560 17. 147

Find the HCF of each of the following groups:

18. 4, 12, 16 19. 8, 17, 36 20. 15, 45, 105
21. 48, 64, 16 22. 24, 36, 54 23. 34, 85, 119

Find the LCM of each of the following groups:

24. 3, 6, 8, 12 25. 15, 27, 135 26. 7, 21, 28, 42
27. 6, 9, 15, 18 28. 4, 8, 12, 18 29. 14, 21, 42

1.2 Associative laws

The terms of an expression may be grouped in any way.

$8 + 4 - 3 = 9$

The above expression may be written as:

$(8 + 4) - 3 = 12 - 3 = 9$

or alternatively as:

$8 + (4 - 3) = 8 + 1 = 9$

without affecting the result. This is the *associative law of addition and subtraction.*

Similarly, the processes of multiplication and division may be treated as follows.

$3 \times 6 \div 2 = 9$

The result may be achieved by:

$(3 \times 6) \div 2 = 18 \div 2 = 9$

or alternatively by:

$3 \times (6 \div 2) = 3 \times 3 = 9$

This is the *associative law of multiplication and division.*

1.3 Commutative laws

Additions and subtractions may be made in any order.

$6 + 9 - 5 = 10$, $9 - 5 + 6 = 10$ and $-5 + 9 + 6 = 10$
$\therefore 6 + 9 - 5 = 9 - 5 + 6 = -5 + 9 + 6$

This is the *commutative law of addition and subtraction.*

Similarly, the order in which numbers are multiplied or divided has no effect on the result.

$5 \times 9 \div 3 = 15$, $9 \div 3 \times 5 = 15$ and $5 \div 3 \times 9 = 15$
$\therefore 5 \times 9 \div 3 = 9 \div 3 \times 5 = 5 \div 3 \times 9$

This is the *commutative law of multiplication and division.*

1.4 Distributive laws

The use of brackets to group together numbers to be added or subtracted is significant when the group is to be multiplied by another number. Consider the expression $5(3 + 7)$. Each of the numbers in the bracket is to be multiplied by the number outside.

i.e. $5(3 + 7) = 15 + 35$

This is the *distributive law of multiplication.*

Extend this idea to the expression $\frac{8+6}{2}$. The denominator may be distributed to each of the terms in the numerator to give:

$\frac{8}{2} + \frac{6}{2} = 4 + 3 = 7$

or alternatively as:

$\frac{1}{2}(8 + 6) = 4 + 3 = 7$

The *distributive law of division* is therefore identical to the distributive law of multiplication.

1.5 Vulgar fractions

A *fraction* is the name given to a quantity that is *part of a whole.* It follows that a fraction must be less than unity. Vulgar fractions are written in the form of two numbers, one vertically below the other. separated by a line called the *bar* of the fraction. In print the bar is often an oblique stroke. For example, $\frac{3}{4}$ or 3/4 indicates that a unit has been divided into *four parts*, and *three* of them are being used.

The top number is called the *numerator* (it enumerates the number of parts) and the bottom number is the *denominator* (it names the parts).

When the numerator is greater than the denominator, the fraction is obviously not less than unity and therefore not a proper fraction. Such fractions are called *improper fractions.*

e.g. $\frac{12}{5}$ is an improper fraction

If the denominator is divided into the numerator, the result is either an integer (if there is no remainder) or a combination of integer and fraction, a *mixed number.*

e.g. $\frac{12}{5} = 2\frac{2}{5}$

Conversely, mixed numbers may be converted into improper fractions by multiplying the integral part by the denominator of the fractional part and adding the numerator to the product, this giving the numerator of the improper fraction.

e.g. $3\frac{2}{3} = \frac{11}{3}$

1.6 Cancelling

Cancelling is the process of simplifying a fraction by dividing both the numerator and denominator by a common factor, preferably their HCF.

e.g. $\frac{15}{25} = \frac{3 \times 5}{5 \times 5} = \frac{3}{5}$ (since $\frac{5}{5} = 1$)

This would normally appear in calculations as

$$\frac{\not{1}\not{5}\,3}{\not{2}\not{5}\,5}$$

It is also quite legitimate, and often convenient, to reverse the process and multiply both numerator and denominator by a number.

e.g. $\frac{2}{5} = \frac{2 \times 2}{5 \times 2} = \frac{4}{10}$

1.7 Zero

Zero cannot be recognised as a factor of any number since nought, nil or nothing taken any number of times can only produce a zero result. It follows that *division or cancellation by zero is not permissible.*

1.8 Manipulating fractions

ADDITION AND SUBTRACTION

Proceed in the following manner, keeping the whole of the expression in each completed line:

i) collect whole numbers together,
ii) reduce fractions to a common denominator (the LCM of the denominators),
iii) collect the numerators together over the common denominator,
iv) simplify the fraction and combine it with the whole number.

Example

$4\frac{2}{3} + 3\frac{1}{4} = 7 + \frac{2}{3} + \frac{1}{4}$ (collecting whole numbers)

$= 7 + \frac{8 + 3}{12}$ (the LCM of 3 and 4 is 12)

$= \underline{7\frac{11}{12}}$

Example

$4\frac{3}{5} - 3\frac{1}{15} + 1\frac{7}{10} = 2 + \frac{3}{5} - \frac{1}{15} + \frac{7}{10}$ (collecting whole numbers)

$= 2 + \dfrac{18 - 2 + 21}{30}$ (the LCM of 5, 10 and 15 is 30)

$= 2 + \frac{37}{30} = 2 + 1\frac{7}{30} = \underline{3\frac{7}{30}}$

Note that the numbers must be taken progressively. In the case of the whole numbers, the 3 is subtracted from the 4 and then the 1 added. It would have been incorrect to add the 3 and the 1 together first and then subtract the result from 4. The same applies to the numerators of the fractions.

MULTIPLICATION

The following procedure should be used:

i) convert mixed numbers to improper fractions,
ii) perform all possible cancelling,
iii) multiply all numerators together and all denominators together,
iv) if the resulting fraction is improper, convert it to a mixed number.

Example

$$4\frac{2}{3} \times 1\frac{5}{6} = \frac{{}^{7}\cancel{14}}{3} \times \frac{11}{\cancel{6}_{3}} = \frac{77}{9} = \underline{8\frac{5}{9}}$$

Note the cancelling by 2 at the second stage.

The use of the word ‘of’ is common when fractions are used, and it means ‘multiply’.

e.g. $\frac{1}{2}$ of $8 = \frac{1}{2} \times 8 = 4$

Example

$$\frac{4}{5} \text{ of } 18\frac{1}{2} = \frac{{}^{2}\cancel{4}}{5} \times \frac{37}{\cancel{2}_{1}} = \frac{74}{5} = \underline{14\frac{4}{5}}$$

Note the cancelling by 2 at the second stage.

Example

$$3\frac{3}{8} \times 1\frac{2}{3} \times 4\frac{4}{15} = \frac{{}^{3}\cancel{9}\cancel{27}}{\cancel{8}_{1}} \times \frac{{}^{1}\cancel{5}}{\cancel{3}_{1}} \times \frac{{}^{8}\cancel{64}}{\cancel{15}\cancel{3}_{1}} = \underline{24}$$

In this example *secondary cancelling* is used. The 3 resulting from dividing 5 into the fifteen is divided into the 9 resulting from dividing 3 into the 27.

DIVISION

Use the following procedure:

i) convert mixed numbers into improper fractions,

ii) invert each divisor and proceed with multiplication.

Example $4\frac{1}{2} \div \frac{1}{2}$. Since each unit contains two halves, $4\frac{1}{2}$ contains $4\frac{1}{2} \times 2 = 9$ halves.

i.e. $\dfrac{4\frac{1}{2}}{\frac{1}{2}} = \dfrac{9}{2} \times \dfrac{2}{1} = \underline{9}$

Example

$$6\tfrac{1}{4} \div 1\tfrac{1}{2} = \frac{25}{4} \div \frac{3}{2} = \frac{25}{\cancel{4}_2} \times \frac{\cancel{2}^1}{3} = \frac{25}{6}. = \underline{4\tfrac{1}{6}}$$

1.9 Ratio and proportion

A vulgar fraction may be described as a *ratio* that expresses the relationship in value between the numerator and the denominator. The fraction $\frac{3}{5}$ has this relationship as '3 to 5', which may be written 3 : 5. Another pair of values in the same ratio are 6 and 10 since $\frac{6}{10} = \frac{3}{5}$. This statement of equality may be expressed in words as '6 is to 10 as 3 is to 5'. The ratio is the same in both cases and the four numbers are said to be '*in proportion*'.

Example The construction of a concrete floor requires 50 kg of cement to each $1\frac{1}{4}$ m^2 of floor laid. How much cement is required for such a floor covering an area of 300 m^2?

The ratio of cement to floor area is 50 : $1\frac{1}{4}$. Therefore, to keep the same proportions:

$$\frac{\text{Required cement}}{300} = \frac{50}{1\frac{1}{4}}$$

Multiply both sides by 300:

$$300 \times \frac{\text{Required cement}}{300} = \frac{50}{1\frac{1}{4}} \times 300$$

$$\text{Required cement} = {}^{10}\cancel{50} \times 300 \times \frac{4}{\cancel{5}_1}$$

$$= \underline{12\,000\,\text{kg}}$$

1.10 Extended operations with fractions

As indicated in section 1.8, care is needed in manipulating numbers with reference to their signs.

Example

$2\frac{1}{2} + 3\frac{7}{8} - 1\frac{2}{3} + 2\frac{1}{6}$

Consider the whole numbers. There is a danger that the beginner may add the 2 and 3, obtaining 5, and then subtract the sum of 1 and 2. The result would then be 2, which is not correct because the 2 would have been subtracted instead of added. The correct result is 6. Continuing the evaluation we have:

$$6 + \frac{12 + 21 - 16 + 4}{24} = 6\frac{21}{24} = \underline{6\frac{7}{8}}$$

When multiplying and dividing fractions, remember that *all the divisors* in an expression should be inverted and changed to multipliers.

Example

$$2\frac{5}{16} \div \frac{5}{8} \times 2\frac{1}{3} \div 2\frac{4}{5} = \frac{37}{16} \div \frac{5}{8} \times \frac{7}{3} \div \frac{14}{5}$$

$$= \frac{37}{\cancel{16}_2} \times \frac{{}^1\cancel{8}}{\cancel{5}_1} \times \frac{{}^1\cancel{7}}{3} \times \frac{{}^1\cancel{5}}{\cancel{14}_2} = \frac{37}{12} = \underline{3\frac{1}{12}}$$

Note the cancelling by 8, 7 and 5.

1.11 Precedence rules

Mixed operations combining the various manipulative processes require the observation of a certain *order of precedence* if mistakes in the application of signs are to be avoided. The rules to be observed may be summarised as follows. The initial letters provide a popular memory aid, BODMAS:

i) evaluate *brackets*, internal brackets first,
ii) change '*of*' to 'multiply',
iii) invert *divisors* and change to multipliers,
iv) carry out *multiplication*,
v) carry out *addition* and subtraction,
vi) *simplify* the result if necessary.

Example

$$2\frac{2}{3}\left(6\frac{1}{4} - 3\frac{1}{2}\right) + 2\frac{7}{8} = 2\frac{2}{3}\left(3\frac{1-2}{4}\right) + 2\frac{7}{8}$$

$$= 2\frac{2}{3} \times 2\frac{3}{4} + 2\frac{7}{8} = \frac{{}^2\cancel{8}}{3} \times \frac{11}{\cancel{4}_1} + 2\frac{7}{8}$$

$$= \frac{22}{3} + 2\frac{7}{8} = 7\frac{1}{3} + 2\frac{7}{8}$$

$$= 9\frac{8 + 21}{24} = 9\frac{29}{24} = \underline{10\frac{5}{24}}$$

Example

$$[(5\tfrac{1}{4} - 2\tfrac{3}{8}) \div 1\tfrac{1}{2}] + [3\tfrac{1}{4} \div (\tfrac{2}{3} + 1\tfrac{1}{15})]$$

$$= [(3\tfrac{2-3}{8}) \div 1\tfrac{1}{2}] + [3\tfrac{1}{4} \div (1\tfrac{10+1}{15})]$$

$$= (2\tfrac{7}{8} \div 1\tfrac{1}{2}) + (3\tfrac{1}{4} \div 1\tfrac{11}{15})$$

$$= \left(\frac{23}{\cancel{8}\,4} \times \frac{{}^{1}\cancel{2}}{3}\right) + \left(\frac{{}^{1}\cancel{13}}{4} \times \frac{15}{\cancel{26}\,2}\right) \quad \text{(note the cancelling)}$$

$$\Rightarrow \frac{23}{12} + \frac{15}{8}$$

$$= 1\tfrac{11}{12} + 1\tfrac{7}{8}$$

$$= 2\tfrac{22+21}{24} = 2\tfrac{43}{24} = \underline{3\tfrac{19}{24}}$$

Note how the whole expression is carried forward step by step. This is a far better approach than scattering various parts of the work over the paper and bringing them back together afterwards.

Exercise 1b Convert the following to mixed numbers:

1. $\frac{171}{12}$ 2. $\frac{235}{8}$ 3. $\frac{22}{7}$ 4. $\frac{41}{5}$ 5. $\frac{625}{144}$ 6. $\frac{53}{16}$

Convert the following to improper fractions:

7. $5\frac{4}{7}$ 8. $11\frac{3}{5}$ 9. $16\frac{7}{8}$ 10. $10\frac{27}{128}$ 11. $22\frac{13}{15}$ 12. $7\frac{3}{64}$

Evaluate:

13. $6\frac{1}{3} - 3\frac{1}{7} + 4\frac{1}{2}$

14. $2\frac{1}{9} + 3\frac{2}{5} - 1\frac{2}{3} + 4\frac{2}{15}$

15. $7\frac{1}{2} + 3\frac{2}{5} + 4\frac{11}{15}$

16. $5\frac{11}{16} - \frac{3}{4} + 2\frac{5}{8} - 1\frac{11}{32}$

17. $3\frac{1}{7} \times 49 \times \frac{1}{2}$

18. $5\frac{5}{8} \div \frac{3}{4}$

19. $5\frac{1}{4} \times 1\frac{2}{3} \div 2\frac{3}{5}$

20. $7\frac{1}{4} \div 1\frac{3}{8} \times 2\frac{3}{16}$

21. $\dfrac{3\frac{2}{3} \times 2\frac{1}{2}}{3\frac{2}{3} - 2\frac{1}{2}}$

22. $2(3\frac{3}{4} - 1\frac{5}{8}) + 3(1\frac{1}{12} + 2\frac{5}{6})$

23. $\dfrac{\frac{3}{4} \text{ of } \frac{7}{9} + 1\frac{3}{7} \div \frac{5}{14}}{3\frac{3}{4} \div 1\frac{1}{2} - \frac{3}{5} \text{ of } 1\frac{7}{8}}$

24. $\dfrac{\frac{2}{9} + \frac{4}{5}(2\frac{1}{4} - 1\frac{2}{5})}{(\frac{7}{8} - \frac{1}{5}) \div (\frac{2}{7} + 1\frac{1}{2})}$

25. $\frac{3}{8}\,[5\frac{1}{4}(6\frac{1}{2} - 3\frac{1}{7}) + \frac{5}{8}(\frac{2}{5} + 1\frac{1}{16})]$

2 THE DECIMAL SYSTEM

2.1 Decimal fractions Usually referred to as *decimals*, these are fractions that would have *multiples of ten* as their denominators if the denominators were written. So long as this is recognised it is unnecessary to write the denominators. A *decimal point* is used to separate the whole number from the fraction in the case of mixed numbers. In the case of proper fractions a nought is placed before the decimal point.

2.2 Conversion of fractions

Examples are:

$0.5 = \frac{5}{10} = \frac{1}{2}$

$0.05 = \frac{5}{100} = \frac{1}{20}$

$0.005 = \frac{5}{1000} = \frac{1}{200}$ and so on

To change a vulgar fraction to a decimal fraction, divide the numerator by the denominator.

$\frac{4}{5} = 0.8$

$\frac{3}{8} = 0.375$

$2\frac{3}{4} = 2.75$

2.3 Recurring decimals

Some vulgar fractions convert to decimals in which successive figures, or groups of figures, repeat.

e.g. $\frac{1}{9} = 0.111\,111\,11$ . . . etc.

This is a *recurring decimal*, and a dot placed over the repeating figure is used to indicate this.

e.g. $\frac{1}{9} = 0.\dot{1}$

If two digits repeat, a dot is placed over each.

e.g. $\frac{9}{11} = 0.818\,181$ . . . etc.

$= 0.\dot{8}\dot{1}$

If a larger group of digits recur, a dot is placed over the first and last digits in the group.

e.g. $\frac{8}{27} = 0.296\,296\,296$ . . . etc.

$= 0.\dot{2}9\dot{6}$

If the division process reaches a stage where there is no remainder, the result is a *finite* or *terminating decimal.*

e.g. $\frac{5}{8} = 0.625$

When the process has no natural end, the result is an *infinite* or *non-terminating decimal.*

e.g. $\frac{1}{7} = 0.\dot{1}42\,85\dot{7}$

2.4 Degree of accuracy

The degree of accuracy required in a solution governs the number of decimal places to which the result is expressed. The result is described as being 'correct to so many decimal places'. In order to make such a correction, the working is carried to one place farther than that required Then, if the last figure is 5 or over, the preceding figure is increased by 1. If the last figure is less than 5, it is rejected.

e.g. 2.758 is 2.76 correct to two decimal places

5.673 is 5.67 correct to two decimal places

2.5 Significant figures

Another method of expressing an approximation is to describe a result as being correct to so many *significant figures*. The procedure is then to carry the calculation to one place beyond the required limit this time including all the digits in the number. The correction is then made as before, discarding the last figure if it is less than 5, or increasing the previous figure by 1 if the last figure is 5 or over.

e.g. 21.87 is 21.9 correct to three significant figures
37 684 is 37 680 correct to four significant figures
554 679 is 555 000 correct to three significant figures

2.6 Fractions as percentages

The words *per cent* mean *per hundred* or *out of a hundred*, and are denoted by the symbol %. Thus 10% means $\frac{10}{100}$, which cancels to $\frac{1}{10}$ and which is 0.1 as a decimal fraction. Similarly 25% means $\frac{25}{100}$, which cancels to $\frac{1}{4}$ as a vulgar fraction and is 0.25 as a decimal.

Vulgar fractions may be converted to percentages simply by multiplying by 100 and adding the % sign. Since this is equivalent to multiplying by $\frac{100}{100}$ the value remains unchanged.

e.g. $\frac{1}{8} = \frac{1}{8} \times 100\% = \frac{100}{8}\% = 12\frac{1}{2}\%$ or 12.5%

Decimal fractions are converted to percentages in the same way.

e.g. $0.25 = 0.25 \times 100\% = 25\%$

$0.375 = 0.375 \times 100\% = 37.5\%$

Exercise 2a

Convert the following vulgar fractions to decimals, correct to three decimal places if non-terminating:

1. $\frac{21}{64}$	2. $\frac{2}{3}$	3. $\frac{5}{8}$
4. $\frac{1}{4}$	5. $\frac{7}{16}$	6. $\frac{27}{40}$
7. $3\frac{21}{27}$	8. $\frac{9}{20}$	9. $1\frac{14}{25}$

Convert the following decimals to vulgar fractions in their simplest form:

10. 0.875	11. 0.2	12. 0.45
13. 10.36	14. 0.125	15. 6.025

Express the following as recurring decimals:

16. $\frac{5}{6}$	17. $\frac{7}{18}$	18. $\frac{19}{27}$

Convert the following into percentages to the nearest $\frac{1}{2}\%$:

19. $\frac{7}{8}$	20. $\frac{11}{25}$	21. $\frac{31}{32}$
22. $\frac{6}{35}$	23. $\frac{17}{24}$	24. $\frac{127}{150}$
25. 0.736	26. 0.1254	27. 0.268

2.7 Addition and subtraction of decimals

This is carried out under the same rules of sign and precedence as are applied to integers and vulgar fractions. Confusion may be avoided by operating in vertical columns.

Example

173.526 + 84.9 + 2.511 + 0.27

```
173.526
 84.900     Note the filling of blank spaces with
  2.511     noughts to improve the visual effect.
  0.270
-------
261.207   (261.2 correct to four significant figures)
```

Example

5.73 + 18.25 − 4.63 + 8.1 − 17.34

Positive	5.73	Negative	4.63
	18.25		17.34
	8.10		
			21.97
	+32.08		
	−21.97		
	10.11		

Example

(53.86 − 17.75) + 15.37 − (54.325 − 11.63)

1st bracket	53.86	2nd bracket	54.325
	−17.75		−11.630
	36.11		42.695

The expression is now:

36.11 + 15.37 − 42.695
+15.37

51.480
−42.695

8.785 (8.79 correct to two decimal places)

2.8 Multiplication and division of decimals

The number of decimal places in a product is the sum of the number of decimal places in the multiplicand and the multiplier together. Any noughts resulting from the multiplication process must be included.

This is illustrated by considering 0.1 × 0.1. If written in vulgar fractions this is $\frac{1}{10} \times \frac{1}{10} = \frac{1}{100}$. But $\frac{1}{100} = 0.01$, therefore 0.1 × 0.1 = 0.01.

Example Find the product of 16.345 and 7.32, and correct the result to three decimal places.

Since there are five decimal places in the two numbers together, there will be five decimal places in the product. Arrange the multiplier below the multiplicand with their decimal points vertically in line.

	16.345
	7.32
Multiply by the 0.02 (5 places):	0.326 90
Now by the 0.3 (4 places):	4.903 5
Finally by the 7 (3 places):	114.415
Add:	119.645 40

The product is 119.645 correct to three decimal places. Note the importance of including the last nought in the first line when assessing the number of decimal places.

Division of decimals is reduced to straightforward long division by the adjustment of the dividend and divisor until the divisor becomes an integer.

Example Evaluate 16.32 ÷ 0.02.

This may be written as the fraction $\frac{16.32}{0.02}$. Multiply both the numerator and the denominator by 100 (i.e. move all the figures two places to the left). The fraction then becomes $\frac{1632}{2}$, which is still the same ratio. The quotient is 816.

Example 9.375 ÷ 4.76 correct to three significant figures.

$$\frac{9.375}{4.76} = \frac{937.5}{476}$$ (multiplying numerator and denominator by 100)

```
        1.969
476)937.500
     476          (1 x 476)
     461.5        (remainder with 0.5 brought down)
     428.4        (0.9 x 476)
      33.10       (remainder with a nought brought down)
      28.56       (0.06 x 476)
       4.540      (remainder with a nought brought down)
       4.284      (0.009 x 476)
      stop
```

The quotient is 1.97 correct to three significant figures.

Exercise 2b Evaluate the following, giving the result correct to four significant figures:

1. 31.65 (9.876 – 5.43 + 2.175)
2. (81.76 – 52.43) (76.8 + 1.234 – 65.7)
3. 0.25 (43.2 – 42.3) + (5.64 ÷ 1.2)
4. $\dfrac{1.125\ (21.63 - 11.29)}{(1.53 + 2.76 - 0.83)}$

Evaluate, correct to three decimal places:

5. 1.24 x 15.36 6. 47.31 x 2.65 7. 4.7 x 0.06 x 18.54

Find the following quotients, correct to two decimal places:

8. 51.318 ÷ 5.06 9. 3817.95 ÷ 12.5 10. 101.95 ÷ 21.63

Evaluate the following expressions, correct to four significant figures:

11. $\dfrac{61.35 \times 1.23}{1.76 \times 10.31}$ 12. $\dfrac{723.5 \times 0.076}{125.4 \times 1.11}$

3 INDICES

3.1 The meaning of index

When a number is multiplied by itself, the result is called a *power* of the number. The number of factors used to produce the power is indicated in small figures at the top right of the base number, and is called an *index*. The plural of index is *indices.*

e.g. $3 \times 3 = 3^2$

In the above example the *base* number is 3, the product is the *second power of three* and the *index* is 2.

$5 \times 5 \times 5 = 5^3$

This is the *third power of five*, the index is 3.

A number raised to its second power is said to be *squared* and, when raised to its third power, to be *cubed.*

e.g. $1.2^3 = 1.2 \times 1.2 \times 1.2 = 1.728$
$0.75^2 = 0.75 \times 0.75 = 0.5625$
$3.5^5 = 3.5 \times 3.5 \times 3.5 \times 3.5 \times 3.5$
$= 525.22$ correct to two decimal places

Finding a *root* of a number is the reverse process of finding a power. The *square root* (or second root) of a number is the factor which has to be used *twice* to produce the number.

e.g. the square root of 9 is 3, because 3×3 (or 3^2) = 9

$\sqrt{\ }$ is the square root sign (the radical sign)

$\sqrt[3]{\ }$ means cube root

$\sqrt[4]{\ }$ means fourth root, and so on

An alternative method of expressing a root is in the form of a *fractional index*:

Square root is the power of a half

Cube root is the power of a third

Fourth root is the power of a quarter, etc.

e.g. $\sqrt{25} = 25^{\frac{1}{2}} = 5$
$\sqrt[4]{81} = 81^{\frac{1}{4}} = 3$
$\sqrt[5]{32} = 32^{\frac{1}{5}} = 2$

The *reciprocal* of a number is the result of dividing the number into 1. It is also called the *inverse* of the number, since it is obtained by inverting the number.

e.g. the reciprocal of 4 is $\frac{1}{4}$ (0.25 as a decimal)

The reciprocal of a power such as 3^2 is $1/3^2$ but may alternatively be expressed as 3^{-2}. In other words *a negative index indicates a reciprocal value.*

e.g. 10^{-1} means $\frac{1}{10}$ and 6^{-3} means $\frac{1}{6^3}$

3.2 Laws of indices

If the product of a number of powers of the same base is required, a simple law of procedure may be applied.

Suppose the product $3^2 \times 3^4 \times 3$ is required. All the factors are powers of the base number 3. Writing the product as a multiple of 3 we obtain:

$$\begin{aligned} & 3^2 \times 3^4 \times 3^1 \\ &= (3 \times 3) \times (3 \times 3 \times 3 \times 3) \times 3 = \underline{3^7} \end{aligned}$$

The product is shown to be the seventh power of three, and the index 7 is the sum of the indices 2, 4 and 1. The law is: *to multiply, add the indices.* It is the first law of indices, and it will be referred to again in the later chapter on algebra. Remember that it is only valid for powers of a *common base.*

e.g. $5^3 \times 5^6 = 5^{3+6} = \underline{5^9}$

$4^2 \times 4^5 \times 4 = 4^{2+5+1} = \underline{4^8}$

Division of powers to a common base may also be shown to obey a law, the second law of indices. Consider the expression:

$$\frac{2^8}{2^5}$$

Expressing both as multiples of 2, we obtain:

$$\frac{2 \times 2 \times 2 \times \cancel{2} \times \cancel{2} \times \cancel{2} \times \cancel{2} \times \cancel{2}}{\cancel{2} \times \cancel{2} \times \cancel{2} \times \cancel{2} \times \cancel{2}}$$

Five cancelling processes reduces the value to $2 \times 2 \times 2 = 2^3$. The index 3 is the difference between the indices of the dividend and divisor, 8 and 5 respectively.

i.e. $2^8 \div 2^5 = 2^{8-5} = \underline{2^3}$

The law is: *to divide, subtract the indices.* Like the first law, it is only valid for powers of a *common base.*

e.g. $7^8 \div 7^3 = 7^{8-3} = \underline{7^5}$

$5^6 \div 5^2 = 5^{6-2} = \underline{5^4}$

The two laws can be combined.

e.g. $\dfrac{3^7 \times 3^5}{3^9} = 3^{7+5-9} = \underline{3^3}$

When a power is raised to a power, the result may be shown to be given by a third law of indices. Consider the power of a power, $(2^3)^2$.

$$\begin{aligned} 2^3 &= 2 \times 2 \times 2 \\ \therefore (2^3)^2 &= (2 \times 2 \times 2) \times (2 \times 2 \times 2) = \underline{2^6} \end{aligned}$$

The result is the sixth power of the base 2. The index 6 is the product of the indices 2 and 3. The law is: *to raise a power to a power, multiply the indices.*

e.g. $(4^5)^3 = 4^{5 \times 3} = \underline{4^{15}}$

$(10^3)^2 = 10^{3 \times 2} = \underline{10^6}$

The following examples show some simple calculations carried out by application of the laws of indices. It will be seen in a later chapter that these laws are of prime importance in the application of logarithms.

Example Evaluate:

$$\frac{16 \times 256}{128}$$

Notice that $16 = 2^4$
$256 = 2^8$
$128 = 2^7$

The whole expression may therefore be written in terms of powers of 2 as:

$$\frac{2^4 \times 2^8}{2^7}$$

$$= 2^{4+8-7} = 2^5 = \underline{32}$$

Example Evaluate:

$$\frac{(\sqrt{3})^4 \times 243}{27}$$

Notice that $\sqrt{3} = 3^{\frac{1}{2}}$
$243 = 3^5$
$27 = 3^3$

The expression then becomes:

$$\frac{(3^{\frac{1}{2}})^4 \times 3^5}{3^3} = \frac{3^2 \times 3^5}{3^3}$$

$$= 3^{2+5-3} = 3^4 = \underline{81}$$

3.3 Zero index and negative indices

Zero index (the power of nought) will produce *unit* value when applied to any base number.

Consider any power of any base, say 5^2. If we divide it by itself the quotient must be 1.

i.e. $\frac{5^2}{5^2} = 1$

If we now apply the second law of indices (to divide, subtract the indices) to the same operation we have:

$$\frac{5^2}{5^2} = 5^{2-2} = 5^0$$

Thus we have shown that $5^0 = 1$. The result will be the same whatever the base and power used.

A similar manipulation may be used to shown that inverting a power changes the sign of the index (see section 3.1). Consider the power 5^4. If we invert it we have its reciprocal $1/5^4$. But we have shown above that $5^0 = 1$. Therefore:

$$\frac{1}{5^4} = \frac{5^0}{5^4}$$

Applying the second law of indices:

$$\frac{5^0}{5^4} = 5^{0-4} = 5^{-4}$$

i.e. $\underline{\frac{1}{5^4} = 5^{-4}}$

(Invert the power, change the sign of its index.)

Exercise 3 Find the value of:

1. 2^7 2. 5^3 3. 6^3

4. $(0.8)^2$ 5. $(1.1)^2$ 6. 4^5

Evaluate:

7. $\sqrt{81}$ 8. $\sqrt[3]{343}$ 9. $\sqrt[4]{16}$

10. $\sqrt[5]{243}$ 11. $(64)^{\frac{1}{2}}$ 12. $(125)^{\frac{1}{3}}$

Use the laws of indices to evaluate:

13. $3^2 \times 3^4 \times 3$ 14. $2^5 \times 2^2 \times 2$

15. $\frac{27 \times 81}{243}$ (base 3) 16. $\frac{8 \times 64 \times 256}{16 \times 128}$ (base 2)

17. $\frac{49 \times 343 \times 7}{(\sqrt{7})^6}$ (base 7) 18. $\frac{(\sqrt{25})^4 \times \sqrt[3]{125}}{25}$ (base 5)

Find the value of:

19. $\frac{1}{12^{-2}}$ 20. $\frac{8^{-1}}{8^{-2}}$ 21. $\frac{1}{\sqrt{8} \times \sqrt{32}}$

22. $(3^{1.2})^5$ 23. $(2^{0.8})^5$ 24. $16 \times (\sqrt{256})^{\frac{1}{2}}$

4 STANDARD FORM AND BINARY FORM

4.1 Standard form

Our system of numbers in common use is a *denary system*, which means *in the scale of ten.* Very large numbers and very small numbers tend to be cumbersome.

e.g. 35 760 000 and 0.000 012

A convenient way of presenting them more compactly is in terms of powers of ten. They are then said to be in *standard form.*

Take 35 760 000 as an example. We may write it as:

$3.756 \times 10\,000\,000$

but 10 000 000 is 10^7.

$\therefore$ 35 760 000 = $\underline{3.576 \times 10^7}$

This is standard form. Applying a similar process to 0.000 012, we may write it as:

$1.2 \times 0.000\,01$

Now, 0.000 01 is 1/100 000 or $1/10^5$, which is 10^{-5} (see section 3.3).

$\therefore$ 0.000 012 = $\underline{1.2 \times 10^{-5}}$

Standard form consists of a number in decimal form between 1 and 10 (with only one digit before the decimal point) multiplied by a power of ten to produce the required value. The index of the power of ten is the *exponent.*

A rule for converting denary numbers to standard form is: move the decimal point the number of places required to produce the number (one digit before the decimal point) and multiply by 10 to the power of the number of places moved. If the decimal point has to be moved to the *left* the index is *positive*, if to the *right*, *negative.*

e.g. 86 700 000 = 8.67×10^7

1 233 000 = 1.233×10^6

84 900 = 8.49×10^4

0.000 000 22 = 2.2×10^{-7}

0.001 75 = 1.75×10^{-3}

4.2 Conversion from standard form of denary

Reversal of the above rule will enable you to convert from standard form to a denary number.

e.g. 9.53×10^6 = 9 530 000

1.725×10^4 = 17 250

5.61×10^{-4} = 0.000 561

3.076×10^{-3} = 0.003 076

As the examples show, the index applied to ten (the exponent) gives the number of places through which the decimal point must be moved. If the exponent is *positive* the decimal point is moved to the *right*, if *negative* to the *left*. Noughts are inserted in any spaces left blank by movement of the point.

4.3 Manipulating numbers in standard form

The addition and subtraction of numbers expressed in standard form is carried out after all the numbers have been adjusted to have a common exponent equal to the *greatest exponent* in the group.

Example

$(1.29 \times 10^6) + (3.57 \times 10^4)$

$= (1.29 \times 10^6) + (0.0357 \times 10^6)$

$= \underline{1.3257 \times 10^6}$

Note that in the second bracket 3.57 has been divided by 10^2, and 10^4 has been multiplied by 10^2. The value within the bracket therefore remains unaltered but the exponent 10^6 is now common to both numbers.

Example

$$(7.35 \times 10^7) - (8.27 \times 10^6)$$
$$= (7.35 \times 10^7) - (0.827 \times 10^7)$$
$$= \underline{6.523 \times 10^7}$$

Example

$$(1.76 \times 10^5) + (2.534 \times 10^6) - (3.7 \times 10^4)$$
$$= (0.176 \times 10^6) + (2.534 \times 10^6) - (0.037 \times 10^6)$$
$$= \underline{2.673 \times 10^6}$$

Example

$$(1.75 \times 10^{-4}) + (2.63 \times 10^{-2})$$
$$= (0.0175 \times 10^{-2}) + (2.63 \times 10^{-2})$$
$$= \underline{2.6475 \times 10^{-2}}$$

Note that, in the case of negative exponents, the greatest index is the one with the lowest numeral, i.e. in the above example –2 is greater than –4.

Exercise 4a Express the following numbers in standard form:

1.	7 329	2.	14 900	3.	325 000
4.	980	5.	0.006 3	6.	0.000 002 7
7.	0.000 8	8.	271 000	9.	0.014 9

Convert the following from standard form to denary numbers:

10.	8.2×10^6	11.	7.9×10^2	12.	1.03×10^5
13.	2.5×10^7	14.	3.2×10^{-3}	15.	1.8×10^{-6}
16.	6.5×10^{-4}	17.	1.01×10^{-5}	18.	5.81×10^{-7}

Evaluate the following in standard form:

19. $(1.85 \times 10^6) + (2.7 \times 10^4)$
20. $(3.65 \times 10^5) + (1.3 \times 10^3)$
21. $(2.83 \times 10^5) - (1.8 \times 10^4)$
22. $(5.7 \times 10^4) - (4 \times 10^3)$
23. $(4.8 \times 10^6) + (1.7 \times 10^5) - (3.2 \times 10^4)$
24. $(7.25 \times 10^7) - (8.6 \times 10^6) - (4.3 \times 10^4)$

4.4 Binary form

In the *binary system* of numbers the *scale of two* is the basis, instead of the scale of ten as in denary. If we arrange a typical series of columns in the denary system followed by a similar series in the binary system, we see that the denary columns progress in *powers of ten* from the units column, but the binary columns progress in *powers of two*.

Denary	thousands	hundreds	tens	units
	10^3	10^2	10^1	10^0
Number:	1	1	1	1
Binary	eights	fours	twos	units
	2^3	2^2	2^1	2^0
Number:	1	1	1	1

Consider the number 1 111. In denary it is one thousand, one hundred and eleven, being made up of:

1 in the 'thousands' column, $1 \times 10^3 = 1\,000$

1 in the 'hundreds' column, $1 \times 10^2 = 100$

1 in the 'tens' column, $1 \times 10^1 = 10$

and 1 in the 'units' column, $1 \times 10^0 = 1$

Now consider the same number in the binary system. It is made up of:

1 in the 'eights' column, $1 \times 2^3 = 8$

1 in the 'fours' column, $1 \times 2^2 = 4$

1 in the 'twos' column, $1 \times 2^1 = 2$

and 1 in the 'units' column, $1 \times 2^0 = 1$

Adding the final column: $\underline{15}$

In other words, 1111 in binary has the value 15 in denary. Taking another example, whereas in the denary system 111 means:

$$10^2 + 10^1 + 10^0$$
$$= 100 + 10 + 1 = \underline{111}$$

in the binary system it means:

$$2^2 + 2^1 + 2^0$$
$$= 4 + 2 + 1 = \underline{7}$$

CONVERSION FROM DENARY TO BINARY

Consider the integer 28. Arrange it as the sum of descending powers of 2, continuing to the units column (i.e. 2^0) and putting noughts in any unused columns.

$$28 = 16 + 8 + 4 + 0 + 0$$
$$= 2^4 + 2^3 + 2^2 + 0 + 0$$
$$= 1 \quad 1 \quad 1 \quad 0 \quad 0 \text{ binary}$$

i.e. $\underline{\text{28 denary} = \text{11100 binary}}$

Note that noughts were placed in the 'empty' 2^1 and 2^0 columns. Another example:

$$\begin{aligned} 17 &= 16 + 1 \\ &= 2^4 + 0 + 0 + 0 + 2 \\ &= 1 \quad 0 \quad 0 \quad 0 \quad 1 \text{ binary} \end{aligned}$$

i.e. 17 denary = 10001 binary

Note again the placing of noughts in the empty columns.

4.5 Conversion from binary to denary

This is simply a reversal of the above process.

$$\begin{aligned} 1010 \text{ binary} &= 2^3 + 0 + 2^1 + 0 \\ &= 8 \qquad + 2 \\ &= 10 \end{aligned}$$

1010 binary = 10 denary

Note that the 2^2 and 2^0 columns are empty. Another example:

$$\begin{aligned} 10101 \text{ binary} &= 2^4 + 0 + 2^2 + 0 + 2^0 \\ &= 16 \qquad + 4 \qquad + 1 \\ &= 21 \end{aligned}$$

10101 binary = 21 denary

4.6 Addition of binary numbers

The addition of binary numbers is carried out by arranging in vertical columns and adding each column, working from the units column. We carry 1 for every 2 in the sum, writing the remaining 1 or a nought as the case may be.

Example Add the binary numbers 111 and 1011.

```
  111
 1011
-----
10010  (see explanation below)
```

Add the units column. The result is 2, which is 1 to carry and 0 to write down. Add the next column and the 1 carried. The result is 3, 2 of which gives 1 to carry, the remaining 1 is written down. The third column contains 1, which, with the 1 carried, gives 2. This means 1 to carry and 0 to write down. The final column is similar, containing only 1, which with the 1 carried gives 2, which in turn is 1 to carry to the fifth column and 0 to write down.

As an exercise, suppose we check the above addition through denary conversion. If we convert 111 binary and 1011 binary to denary numbers, add them together and then convert their sum back to binary, we should get the same result.

$$\begin{aligned} 111 \text{ binary} &= 2^2 + 2^1 + 2^0 \\ &= 4 + 2 + 1 \\ &= 7 \text{ denary} \end{aligned}$$

$$1011 \text{ binary} = 2^3 + 0 + 2^1 + 2^0$$
$$= 8 + 2 + 1$$
$$= \underline{11 \text{ denary}}$$

Add these together:

$$7 + 11 = 18$$

Convert back to binary:

$$18 = 16 + 2$$
$$= 2^4 + 0 + 0 + 2^1 + 0$$
$$= \underline{1 \quad 0 \quad 0 \quad 1 \quad 0}$$

which was the binary sum found above.

Example

```
  1011
  1101
 11001
   101
------
110110
```

110110 (see explanation below)

Units column = 4, which means 2 to carry, 0 to write.
Next column plus the 2 carried = 3, which means 1 to carry, 1 to write.
Third column plus the 1 carried = 3 again.
Fourth column plus the 1 carried = 4; carry 2, write 0.
The fifth column plus 2 carried = 3; write 0, and write 1 in the sixth column.
Try checking this in the way of the previous example.

4.7 Application of the binary system

Whereas in denary numbers the greatest digit in any column is 9, in binary the greatest digit is 1. In fact only 1 and 0 are used. This makes the binary system very useful in computers, where electronic operation can produce a 'switched on' or a 'switched off' state. The 'on-off' process can be operated with tremendous speed, and if we read 1 for 'on' and 0 for 'off' we can operate binary arithmetic extremely quickly. In fact an enormous number of operations can be completed in a split second.

Exercise 4b

Convert the following denary numbers to the binary system:

1. 5	2. 26	3. 3
4. 14	5. 18	6. 38
7. 17	8. 51	9. 63

Convert the following binary numbers to the denary (decimal) system:

10. 101	11. 1101	12. 10101
13. 111011	14. 1010111	15. 11
16. 1001	17. 11011	18. 10

Evaluate the following in the binary system, and check the results by conversion to denary before addition and then converting the sum back to binary:

19.	1101 + 1010	20.	11011 + 110 + 100
21.	11101 + 1011	22.	110111 + 11111 + 10
23.	10110 − 1011	24.	1011 − 101
25.	10010 − 101 + 110	26.	11111 − 1011 − 101

SUMMARY OF A

Integer: a whole number.
Digit: a single-figure integer.
Signs: *Like signs* produce *positive* results, *unlike signs* produce *negative* results.

Arithmetical processes:

Process	*Result*
Addition	Sum
Subtraction	Difference
Multiplication	Product
Division	Quotient

Rules of precedence:

B brackets
O of
D division
M multiplication
A addition and subtraction
S simplification

Factors: components of a product.
HCF: the greatest factor common to a group.
LCM: the least multiple common to a group.
Prime number: one without factors (excepting 1).

Vulgar fraction: $\dfrac{\text{numerator}}{\text{denominator}}$

Recurring decimals: $\frac{1}{3} = 0.\dot{3}$, $\frac{1}{11} = 0.\dot{0}\dot{9}$, $\frac{1}{7} = 0.\dot{1}42857$
Correction to decimal places and significant figures: proceed to one place beyond that required. If the last figure is five or more, add 1 to the preceding figure; if the last figure is less than 5, ignore it.
Conversion of a fraction to a percentage: multiply by 100%.
Multiplication of decimals: the number of decimal places in the product is the sum of the number of places in the multiplicand and multiplier together.
Index: the indicator of a power, i.e. the number of equal factors in a product; e.g. $2^3 = 8$. The index is 3.
Root: a fractional index; e.g. $\sqrt{4} = 4^{\frac{1}{2}} = 2$.
Negative index: indicates a reciprocal, i.e. $2^{-2} = 1/2^2 = 1/4$.
Zero index: any number to the power of nought is unity.
Laws of indices:
1) to multiply add the indices,
2) to divide subtract the indices,
3) to raise a power to a power multiply the indices.
Standard form: the expression of numbers as multiples of powers of 10.
Denary system: numbers in the scale of ten (our normal system).
Binary system: numbers in the scale of two (useful in computing).

REVISION EXERCISE A

1. Find as many sets of factors as you can for each of the following numbers:

a) 91 b) 211 c) 121 d) 45

2. Which of the following are prime numbers?

a) 43 b) 133 c) 171 d) 83

3. What are the prime factors of the following numbers?

a) 27 b) 84 c) 217 d) 39

4. Find the HCF and the LCM of each of the following groups:

a) 8, 14, 16, 18 b) 32, 48, 64

c) 5, 25, 35, 40 d) 12, 18, 21, 24

5. Simplify the following expressions:

a) $15 + 1\frac{3}{4} - 6\frac{2}{3}$ b) $4\frac{5}{13} + \frac{3}{26} - 2\frac{3}{7} - \frac{1}{21}$

c) $3\frac{1}{4} + 2\frac{2}{5} \times 5\frac{5}{12} + 3\frac{5}{8}$ d) $\frac{5}{16}$ of $4\frac{4}{15} + 1\frac{1}{2}(6\frac{1}{2} -$ $\frac{2}{3}$ of $3\frac{5}{8} + 2\frac{5}{9})$

6. Arrange the following in decreasing order of size:

$\frac{6}{17}$, $\frac{12}{37}$, $\frac{8}{25}$, $\frac{17}{53}$

7. Express the following fractions as decimals:

a) $\frac{19}{25}$ b) $\frac{6}{125}$ c) $\frac{27}{64}$ d) $\frac{13}{16}$

8. Express the following decimals as vulgar fractions in their simplest terms:

a) 0.055 b) 0.3125 c) 0.0625 d) 0.0875

9. Simplify the following:

a) $52.725 + 69.83 - 42.7 + 34.25$

b) $27.35 + 2.46\,(11.52 - 4.75 + 3.26)$

c) $36.54 \div 3.6 \times 5$

d) $$\frac{(3.2 \times 0.004 - 1.111)\,0.25}{(4 \div 0.2) - 17.907}$$

10. Evaluate the following powers:

a) $(1\frac{3}{4})^2$ b) $(\frac{2}{3})^3$ c) $(\frac{1}{4})^3$ d) $(7 - 5)^2$

11. Find the value of the following roots:

a) $\sqrt{121}$ b) $\sqrt{\frac{9}{64}}$ c) $\sqrt[3]{27}$ d) $\sqrt{(16 + 9)}$

12. Simplify the following using the laws of indices:

a) $\dfrac{(\sqrt{3})^3 \times 81}{243}$ b) $16^{\frac{1}{2}} \times 2^{-2}$ c) $\dfrac{5 \times 25^{-1} \times 625}{125 \times 5^2}$

13. Express the following with positive indices:

a) $4^{-\frac{1}{2}}$ b) $1/8^{-\frac{2}{3}}$ c) 1^{-2} d) 99^0

14. Find the value of:

a) $7^{-2}/7^{-1}$ b) $1/8^{-2}$ c) $(3^{0.6})^5$

15. Express the following numbers in standard form:

a) 4 372 000 b) 6 897 c) 324 700

16. Convert the following to denary numbers:

a) 6.3×10^6 b) 5.95×10^{-3} c) 4.7×10^4

17. Evaluate the following in standard form:

a) $(2.93 \times 10^6 - 1.54 \times 10^4)$

b) $(7.92 \times 10^5 + 3.86 \times 10^6)$

c) $(7.25 \times 10^5) - (4.73 \times 10^3) + (5.64 \times 10^4) - (3.2 \times 10^5)$

18. Convert the following denary numbers to the binary system:

a) 57 b) 23 c) 35 d) 41

19. Convert the following binary numbers to the denary system:

a) 10011 b) 11001 c) 101001

20. Evaluate the following in the binary system:

a) 10011 + 101 + 10 b) 11011 − 111 + 101

c) 101101 − 11110 + 1011 d) 11101 + 101 + 110

B Calculations

This chapter deals with:
Degree of accuracy, significant figures, approximations
Validity of results
Mathematical tables; squares, square roots and reciprocals
Logarithms as indices, use of logarithms
The slide rule
Calculating machines

5 LIMITS

5.1 Degree of accuracy

Precise solutions to practical problems are not always possible or necessary. *Approximate* results should be to a *degree of accuracy* appropriate to the situation, and this will usually be indicated by the basic data available. It is reasonable to present the solution to the same degree of accuracy as the given information. If this varies the result should be given to one decimal place, or significant figure, more than the least accurate value presented.

Examples of degree of accuracy may be quoted, but one cannot make a final decision on such matters until a specific problem is encountered. For instance, carpentry in roof construction may be quite safely measured to the nearest 2 mm or 3 mm, whereas joinery dimensions should be accurate to 0.5 mm or less. In brickwork, accuracy to the nearest 4 mm or 5 mm may suffice. On the other hand the fitting of machine parts in engineering workshops requires great accuracy and 0.005 mm is not an unusual tolerance. When money is involved, most people will insist on accuracy to the nearest $\frac{1}{2}$p, the smallest coin of the realm, although cheques are written to the nearest 1p.

The calculation of the weight of a block of material from its measured dimensions and the material's density will rarely produce exactly the same result as an accurate weighing process. Reasons for the error include such things as inaccurate measurement, rounded edges and corners, out-of-square faces and slight variations in density. Similar inaccuracies arise when the capacity of a vessel is calculated from measured dimensions and then compared with the actual capacity found by filling the vessel.

Example A block of steel measuring 225 mm × 100 mm × 80 mm is found to have a mass of 13.95 kg. Compare this with a calculated mass, using 7 800 kg/m³ as the steel's density.

$$\begin{aligned}\text{Volume of steel} &= \text{length} \times \text{breadth} \times \text{thickness}\\ &= 225 \times 100 \times 80\ \text{mm}^3\\ &= \underline{1.8 \times 10^6\ \text{mm}^3} \quad \text{(standard form)}\\ 1\ \text{m}^3 &= 10^9\ \text{mm}^3 \quad (1\,000 \times 1\,000 \times 1\,000)\\ \therefore\ 7\,800\ \text{kg/m}^3 &= \frac{7\,800}{10^9}\ \text{kg/mm}^3\end{aligned}$$

$$\text{Mass} = \text{volume} \times \text{density}$$

$$\therefore \text{ Mass of steel} = \frac{1.8 \times 10^6 \times 7\,800}{10^9}$$

$$= 1.8 \times 7.8 \text{ kg}$$

$$= \underline{14.04 \text{ kg}}$$

This shows that the actual mass is 90 g less than the calculated value, which could quite safely be approximated to 14 kg.

Example A tank was carefully filled with a measured quantity of water and found to have a capacity of 52 litres. The internal dimensions of the tank were then measured as 600 mm x 350 mm x 250 mm. Calculate the tank's apparent capacity and compare it with the actual capacity (1 m^3 = 1 000 litres).

$$\text{Capacity} = 0.6 \times 0.35 \times 0.25 \text{ m}^3$$

$$= 0.6 \times 0.35 \times 0.25 \times 1\,000 \text{ litres}$$

$$= \underline{52.5 \text{ litres}}$$

The difference is 0.5 litre, roughly a 1% error, and possible explanations are that the sides and bottom of the tank could be slightly concave and internal corners rounded.

Example Excavation for a rainwater soakaway produced 13 m^3 of earth to be carted away. The soakaway was 2 m square on plan and 2.5 m deep, but the soil was known to increase in bulk by 25% when excavated. Compare the actual amount of soil carted away with the quantity estimated by calculation.

Capacity of soakaway	$= 2 \times 2 \times 2.5 \text{ m}^3$
	$= 10 \text{ m}^3$
Add 25% ($\frac{1}{4}$)	2.5 m^3
Estimated quantity	$= \underline{12.5 \text{ m}^3}$

The difference is 0.5 m^3. This type of calculation can never be exact since the bulking of soil is an extremely variable factor. In addition the measurement of the actual quantity carted away could not be precise. Estimates rounded up to the nearest 1 m^3 are adequate in such cases.

5.2 Validity of results

It is surprising how often mistakes in calculations give results that are quite impractical. A little thought about the working of a problem is always worthwhile and often results in errors being avoided. Checking processes are always much more satisfying when no mistakes are found. A common error is the wrong positioning of the decimal point. Another is the mis-reading of signs.

Answers that are not feasible should be carefully scrutinised and, if necessary, rejected. The reasons may not necessarily lie in the calculations. It is possible that incorrect data may have been used.

Apart from errors and omissions, it is possible that there may be more than one solution to a problem. This will become apparent in further studies of algebra and trigonometry, and in such cases the solution appropriate to the basic data must be selected.

5.3 Approximations

Rough estimation of the expected result of a calculation process should not present a lot of difficulty. Rounding off of numbers to integers followed by approximate cancelling is one method of approach. The use of standard form is another.

Example Find an approximate value of:

$$\frac{255 \times 81.7}{36}$$

This is a simple example.

$$\frac{255}{36} \simeq 7$$

$$81.7 \simeq 80$$

$$7 \times 80 = \underline{560}$$

(The value by slide rule is 578.8.)

Example

$$\frac{9.5 \times 389 \times 1\,540}{0.65 \times 8\,400}$$

This time write the numbers in standard form before approximating.

$$= \frac{9.5 \times 3.89 \times 10^2 \times 1.54 \times 10^3}{6.5 \times 10^{-1} \times 8.4 \times 10^3}$$

Approximate, and collect the powers of 10 together:

$$\simeq \frac{1 \times 4 \times 1.5 \times 10 \times 10^2 \times 10^3}{7 \times 8 \times 10^{-1} \times 10^3}$$

$$\simeq 0.1 \times 10^4$$

$$\simeq \underline{1\,000}$$

(The value by slide rule is 1 042.)

Example

$$\frac{(274.5 - 163.7) \times 756}{(128.3 - 95.4)}$$

First obtain a rough evaluation of the brackets:

$$\simeq \frac{110 \times \cancel{756}\,25}{1\cancel{30}} \quad \text{(approximate cancelling)}$$

$\simeq 110 \times 25$
$\simeq \underline{2\,750}$

(The value by slide rule is 2 546.)

With practice, approximations may be made quickly with a minimum of written work. Mental assessment of the likely results of arithmetic processes is a good habit to adopt.

5.4 Checking

On no account should wide variations between approximations and detailed evaluations be ignored. Look for obvious mistakes such as errors in multiples of ten when converting to standard form, errors in cancelling, confusion of signs and so on. Misuse of slide rules, logarithms or calculating machines are other pitfalls to be avoided.

Exercise 5

To what degree of accuracy would you give answers to the following?

1. 7.35×8.64 2. 2.731×5.49 3. $\dfrac{325 \times 840}{1\,580}$

Find approximate values for the following by rounding off and rough cancelling:

4. $\dfrac{19.85 \times 6.34}{3.875}$ 5. $\dfrac{14.93 \times 7.05}{3.6}$

6. $\dfrac{82.54 \times 118.6}{28.75 \times 18.9}$ 7. $\dfrac{3.75 \times 11.8}{1.09 \times 19.5}$

8. $\dfrac{(5 + 16.25) \times (18.7 - 7.9)}{393.6}$ 9. $\dfrac{(27.4 + 11.3)}{(18.25 + 19.5) \times (1.15 + 8.75)}$

Approximate the following after converting to standard form:

10. $\dfrac{315 \times 61.2 \times 11.95}{736}$ 11. $\dfrac{81.65 \times 213}{0.76 \times 15.8}$

12. $\dfrac{19.93(18\,300 + 1\,760)}{391 \times 5\,832}$ 13. $\dfrac{1.414(0.475 + 0.0113)}{(0.625 - 0.076)}$

6 MATHEMATICAL TABLES

Mathematical tables are a tremendous aid to calculation. They contain a complete range of tabulated values of logarithms, antilogarithms, squares, square roots, reciprocals and trigonometric ratios. At this level of study, four-figure tables are adequate. Five-figure tables will be found more useful at a later stage.

6.1 Squares, square roots and reciprocals

SQUARES

An extract from a table of squares of numbers appears below, and the following examples show how the tables are read.

SQUARES

	0	1	2	3	4	5	6	7	8	9	Mean Differences 1	2	3	4	5	6	7	8	9
6.5	42.25	42.38	42.51	42.64	42.77	42.90	43.03	43.16	43.30	43.43	1	3	4	5	7	8	9	10	12
6.6	43.56	43.69	43.82	43.96	44.09	44.22	44.36	44.49	44.62	44.76	1	3	4	5	7	8	9	11	12
6.7	44.89	45.02	45.16	45.29	45.43	45.56	45.70	45.83	45.97	46.10	1	3	4	5	7	8	9	11	12
7.0	49.00	49.14	49.28	49.42	49.56	49.70	49.84	49.98	50.13	50.27	1	3	4	6	7	8	10	11	13
7.1	50.41	50.55	50.69	50.84	50.98	51.12	51.27	51.41	51.55	51.70	1	3	4	6	7	9	10	11	13
7.2	51.84	51.98	52.13	52.27	52.42	52.56	52.71	52.85	53.00	53.14	1	3	4	6	7	9	10	12	13
7.3	53.29	53.44	53.58	53.73	53.88	54.02	54.17	54.32	54.46	54.61	1	3	4	6	7	9	10	12	13

Example Evaluate 7.245^2.

Read across the line '7.2' to the column headed '4' and find 52.42. In the mean differences column headed '5' find a 7. Add this to 52.42 to obtain 52.49.

i.e. $\underline{7.245^2 = 52.49}$

You will notice that the table of squares is printed for numbers from 1 to 10. It may, however, be used for numbers of any magnitude, since the sequence of figures in the squares is not affected by changes in the position of the decimal point in the number. This is illustrated below.

$$2^2 = 4$$
$$20^2 = 400$$
$$200^2 = 40\,000$$

The magnitude of the square varies but the sequence of figures remains the same. The following example illustrates the same point again.

Example Evaluate 6.736^2, 67.36^2 and 673.6^2.

From the tables we read across the '6.7' line and find 45.29 in the '3' column. Under '6' in the mean differences we find an 8, which is added to the 45.29 to give 45.37.

i.e. $\underline{6.736^2 = 45.37}$

If we now consider 67.36^2 we shall obtain the same sequence (4537) in the same way, but since 67.36 is between 60 and 70 its square is between 60^2 (3 600) and 70^2 (4 900).

$\therefore\ 67.36^2 = 4\,537$

Similarly, 673.6 is between 600 and 700 and therefore its square is between 360 000 and 490 000.

$\therefore\ \underline{673.6^2 = 453\,700}$

Note that, since the tables are correct to only four significant figures, two noughts must be added.

SQUARE ROOTS

Square root tables require a little more care in use. You will find that they are in two parts, one for numbers from 1 to 10 and the other for numbers from 10 to 100. This is because a number with an odd number of digits and one with an even number of digits, both having the same sequence of figures, will not have the same sequence of figures in their square roots. This is illustrated in the simple example below.

$$\sqrt{4} = 2$$
$$\sqrt{40} = 6.325$$
$$\sqrt{400} = 20$$
$$\sqrt{4\,000} = 63.25$$
$$\sqrt{40\,000} = 200 \quad \text{etc.}$$

For numbers having an odd number of digits, use the tables for numbers from 1 to 10. For those with an even number of digits, use the tables for numbers from 10 to 100.

For each pair of digits in a number there is one digit in its square root. If the number has an odd number of digits the square root has an extra digit for the odd one.

e.g. $\sqrt{362}$ will have 2 digits
$\sqrt{5\,765}$ will have 2 digits
$\sqrt{76\,890}$ will have 3 digits, etc.

Only the integral part of the number is involved in this rule.

Example Evaluate $\sqrt{7.854}$.

There will be one digit in the integral part of the square root. Using the tables for numbers from 1 to 10 in the extracts printed below, read across the '7.8' line to the '5' column and find 2.802. In the '4' column of the mean differences find a 1, which is added to 2.802 to give 2.803.

$\therefore \sqrt{7.854} = 2.803$

SQUARE ROOTS FROM 1 TO 10

	0	1	2	3	4	5	6	7	8	9	Mean Differences								
											1	2	3	4	5	6	7	8	9
7.5	2.739	2.740	2.742	2.744	2.746	2.748	2.750	2.751	2.753	2.755	0	0	1	1	1	1	1	1	2
7.6	2.757	2.759	2.760	2.762	2.764	2.766	2.768	2.769	2.771	2.773	0	0	1	1	1	1	1	1	2
7.7	2.775	2.777	2.778	2.780	2.782	2.784	2.786	2.787	2.789	2.791	0	0	1	1	1	1	1	1	2
7.8	2.793	2.795	2.796	2.798	2.800	2.802	2.804	2.805	2.807	2.809	0	0	1	1	1	1	1	1	2
7.9	2.811	2.812	2.814	2.816	2.818	2.820	2.821	2.823	2.825	2.827	0	0	1	1	1	1	1	1	2

SQUARE ROOTS FROM 10 TO 100

	0	1	2	3	4	5	6	7	8	9	Mean Differences								
											1	2	3	4	5	6	7	8	9
15	3.873	3.886	3.899	3.912	3.924	3.937	3.950	3.962	3.975	3.987	1	3	4	5	6	8	9	10	11
16	4.000	4.012	4.025	4.037	4.050	4.062	4.074	4.087	4.099	4.111	1	2	4	5	6	7	9	10	11
17	4.123	4.135	4.147	4.159	4.171	4.183	4.195	4.207	4.219	4.231	1	2	4	5	6	7	8	10	11
18	4.243	4.254	4.266	4.278	4.290	4.301	4.313	4.324	4.336	4.347	1	2	3	5	6	7	8	9	10
19	4.359	4.370	4.382	4.393	4.405	4.416	4.427	4.438	4.450	4.461	1	2	3	5	6	7	8	9	10

Example Evaluate $\sqrt{1634}$.

There will be two digits in the square root. As there is an even number of digits in the number, use the tables for numbers from 10 to 100. Read across the '1.6' line to the '3' column, finding 4037. Under the '4' of the mean differences find a 5. Add this to the 4037 to obtain 4042.

$\therefore \quad \underline{\sqrt{1634} = 40.42}$

For square roots of numbers less than 1, the following rules should be observed:

If the figures start immediately after the decimal point, or if there is an even number of noughts after the decimal point, use the tables for numbers from 10 to 100.

If there is an odd number of noughts after the decimal point, use the tables for numbers from 1 to 10.

Examples Follow these examples in the extracts from square root tables printed above:

$\sqrt{0.162\,1} = 0.402\,6$ (10 to 100 tables)
$\sqrt{0.076\,75} = 0.277$ (1 to 10 tables)
$\sqrt{0.001\,843} = 0.042\,93$ (10 to 100 tables)

RECIPROCALS

Tables of reciprocals will be found useful when numbers have to be inverted. (Refer back to section 3.1 for the definition of a reciprocal). It is important to realise that the larger the number the smaller will be its reciprocal.

e.g. reciprocal of 2 $= \frac{1}{2} = 0.5$
reciprocal of 20 $= \frac{1}{20} = 0.05$
reciprocal of 200 $= \frac{1}{200} = 0.005$

For this reason the numbers in the mean differences columns of the reciprocal tables must be subtracted instead of added. A note to this effect appears at the top of the tables, as you will see in the extract below. It is a good idea to underline this note neatly in red.

RECIPROCALS OF NUMBERS. FROM 1 TO 10
(Numbers in difference columns to be subtracted, not added)

											Mean Differences								
	0	1	2	3	4	5	6	7	8	9	1	2	3	4	5	6	7	8	9
5.5	0.1818	1815	1812	1808	1805	1802	1799	1795	1792	1789	0	1	1	1	2	2	2	3	3
5.6	0.1786	1783	1779	1776	1773	1770	1767	1764	1761	1757	0	1	1	1	2	2	2	3	3
5.7	0.1754	1751	1748	1745	1742	1739	1736	1733	1730	1727	0	1	1	1	1	2	2	2	3
5.8	0.1724	1721	1718	1715	1712	1709	1706	1704	1701	1698	0	1	1	1	1	2	2	2	3
5.9	0.1695	1692	1689	1686	1684	1681	1678	1675	1672	1669	0	1	1	1	1	2	2	2	3
6.0	0.1667	1664	1661	1658	1656	1653	1650	1647	1645	1642	0	1	1	1	1	2	2	2	3
6.1	0.1639	1637	1634	1631	1629	1626	1623	1621	1618	1616	0	1	1	1	1	2	2	2	2
6.2	0.1613	1610	1608	1605	1603	1600	1597	1595	1592	1590	0	1	1	1	1	2	2	2	2
6.3	0.1587	1585	1582	1580	1577	1575	1572	1570	1567	1565	0	0	1	1	1	1	2	2	2

Example Find the reciprocal of 5.724.

Read across the '5.7' line to the '2' column to find 0.1748. In the '4' column of the mean differences find a 1, which is *subtracted* from the 0.1748 to give 0.1747.

$\therefore \frac{1}{5.724} = 0.1747$

Although the reciprocal tables are printed for numbers from 1 to 10, they may be used for numbers of any magnitude.

Example Find the reciprocal of 61.35.

$\frac{1}{61.35}$ is between $\frac{1}{10}$ and $\frac{1}{100}$, in other words between 0.1 and 0.01. Using the tables as previously indicated we obtain the sequence of figures 163.

$\therefore \frac{1}{61.35} = 0.0163$

Example Find the reciprocal of 596.8.
$\frac{1}{596.8}$ is between $\frac{1}{100}$ (0.01) and $\frac{1}{1000}$ (0.001). There will therefore be two noughts after the decimal point. From the tables the sequence of figures is 1676.

$\therefore \frac{1}{596.8} = 0.001\,676$

Exercise 6a Use tables of squares, square roots and reciprocals, as appropriate, to evaluate the following:

1. 2.56^2	2. 11.74^2	3. 1.254^2
4. 27.9^2	5. 0.825^2	6. 0.096^2
7. $\sqrt{51.84}$	8. $\sqrt{2.95}$	9. $\sqrt{111.7}$
10. $\sqrt{14.13}$	11. $\sqrt{0.844}$	12. $\sqrt{5\,276}$
13. $\frac{1}{0.463}$	14. $\frac{1}{21.84}$	15. $\frac{1}{3.142}$
16. $\frac{1}{9.81}$	17. $\frac{1}{108}$	18. $\frac{1}{0.0765}$

6.2 Logarithms

In section 3.1 we saw that when a number is raised to a *power*, such as 3^2, the number is called the *base* and the value of the power is called the *index*. We also saw that indices may be used to simplify the processes of multiplication and division, provided the same base is used.

Indices used in this way are called *logarithms*. It is not convenient to use any number chosen at random as a base for logarithms, but the number 10 is a convenient base and logarithms to the base 10 are called *common logarithms*. This is because numbers having the same sequence of figures will have the same decimal part to their logarithms, regardless of their magnitude; consequently tables of logarithms to the base 10 may be printed in a small space.

Because $10^2 = 100$

$$\log_{10} 100 = 2$$

The above statement expressed in words is: '*Because ten squared equals one hundred, the logarithm of one hundred to the base ten is two.*'

The logarithm of a number is an index. It is the index which must be applied to the base to produce the number. The formal definition of a logarithm is usually stated as: *The logarithm of a number to a given base is the index of the power to which the base must be raised to give the number.*

6.3 Reading tables of logarithms

A logarithm has two distinct parts, an *integral* part called the *characteristic* and a *decimal* part called the *mantissa*. Only the mantissa is found in tables of logarithms. The characteristic must be found by reasoning as follows.

$$10\,000 = 10^4 \qquad \therefore \log_{10} 10\,000 = 4$$

$$1\,000 = 10^3 \qquad \therefore \log_{10} 1\,000 = 3$$

$$100 = 10^2 \qquad \therefore \log_{10} 100 = 2$$

$$10 = 10^1 \qquad \therefore \log_{10} 10 = 1$$

$$1 = 10^0 \text{ (see 3.1)} \quad \therefore \log_{10} 1 = 0$$

In all the above cases, since the numbers are integral powers of 10, the logarithms consist of characteristics only, no mantissa. In each case the value of the characteristic is 1 less than the number of digits in the number.

Consider the number 2 365. It is between 1 000 (10^3) and 10 000 (10^4). Its logarithm is therefore between 3 and 4. From the logarithm tables find the mantissa as follows. Read across the '23' line to the '6' column to find 3729. Under the '5' in the mean differences we find a 9, which, added to 3729, gives 3738. This completes the mantissa, and:

mantissa

$$\log_{10} 2\,365 = 3.3738$$

characteristic

which means that $2\,365 = 10^{3.3738}$.

Example Evaluate $\log_{10} 191.5$.

Since there are three digits in the integral part of the number, the characteristic will be 2. From the tables the mantissa is 0.2822,

$$\therefore \log_{10} 191.5 = 2.2822$$

Remember that in *index form* this is written:

$$191.5 = 10^{2.2822}$$

Since we are going to use only logarithms to the base 10, we may write 'log' instead of '$\log_{10}$'.

Although the *mantissa* of common logarithms is *always positive*, numbers less than 1 have logarithms whose *characteristic* is *negative*.

Continuing our previous table of integral powers of 10:

$$1 = 10^0 \qquad \therefore \log_{10} 1 = 0$$

$$0.1 = \frac{1}{10} = \frac{1}{10^1} = 10^{-1} \qquad \therefore \log_{10} 0.1 = -1$$

$$0.01 = \frac{1}{100} = \frac{1}{10^2} = 10^{-2} \qquad \therefore \log_{10} 0.01 = -2$$

$$0.001 = \frac{1}{1\,000} = \frac{1}{10^3} = 10^{-3} \qquad \therefore \log_{10} 0.001 = -3 \quad \text{etc.}$$

The characteristic in each case is 'numerically' one more than the number of noughts after the decimal point, but has a negative sign.

Suppose we require the logarithm of 0.076 43. Since there is one nought after the decimal point, the characteristic is –2. To avoid confusion arising from the fact that this is negative whilst the mantissa will be positive, the minus sign is written above the figure as $\bar{2}$ and called 'bar 2'. From the tables the mantissa is 0.8833.

$\therefore$ $\underline{\log 0.076\,43 = \bar{2}.8833}$

Exercise 6b Write down the logarithms of :

1. 112.9	2. 56.84	3. 7.596
4. 0.123 4	5. 1.040	6. 0.009 73
7. 3 283	8. 11 460	9. 3.87×10^6
10. 4.81×10^4	11. 1.75×10^2	12. 6.54×10^5

6.4 Multiplication and division using logarithms

The use of logarithms for multiplication is an application of the first law of indices: *to multiply, add the indices* (see section 3.2).

Suppose we wish to evaluate 27.93 × 81.54. From the logarithm tables we find that:

$$\log 27.93 = 1.446\,1$$

$$\text{and } \log 81.54 = 1.911\,4$$

This means that:

$$27.93 = 10^{1.4461}$$

$$\text{and } 81.54 = 10^{1.9114}$$

We may therefore write 27.93 × 81.54 as:

$$10^{1.4461} \times 10^{1.9114}$$

$$= 10^{1.4461 + 1.9114}$$

$$= 10^{3.3575} \text{ (by first law of indices)}$$

We now need to express this power of 10 as a number. This we do in tables of *antilogarithms*, which are merely reversed logarithms. Only the mantissa is used in reading the tables, and the result is the sequence of figures 2278.

The number of digits in the integral part of the number (i.e. the position of the decimal point) is found from the characteristic. Since this is 3, there are four figures in the number.

i.e. antilog 3.3575 = 2278

and 27.93 x 81.54 = 2278

The above example illustrates the application of the first law of indices. It is not normally necessary to write out the process in full as powers of 10, and the following method of tabulation is a convenient way of presenting the calculation.

No.	Log
27.93	1.4461
81.54	1.9114
2278	3.3575

antilog

Note. A reverse arrow at the antilogarithm stage is useful for clarity and ease of checking.

Division by logarithms is an application of the second law of indices: *to divide, subtract the indices.*

Example

379.3 ÷ 72.59

(subtract the logs)

= 5.225

No.	Log
379.3	2.5789
72.59	1.8608
5.225	0.7181

Combinations of multiplication and division processes may be set out as shown in the following example.

Example

$$\frac{27.93 \times 138.6}{9.81 \times 17.54}$$

i) Add the logarithms of the two numbers in the numerator,

ii) add the logarithms of the two numbers in the denominator,

iii) subtract the logarithm of the denominator from that of the numerator,

iv) find the antilogarithm.

= 22.5

No.	Log
27.93	1.4461
138.6	2.1418
numerator	3.5879
denominator	2.2357
22.50	1.3522
9.81	0.9917
17.54	1.2440
denominator	2.2357

When numbers less than 1 are involved, having logarithms with negative characteristics, care with addition and subtraction is necessary.

Example

3.142 ÷ 0.771 = 4.07

No.	Log
3.142	0.4972
0.771	1.8871
4.075	0.6101

Note that the process in the characteristic column is the subtraction of $\bar{1}$ from $\bar{1}$ leaving 0 (or 0 from 0 if the borrowed 1 is 'paid back' below).

Example

$$\frac{73.56 \times 0.095}{0.3847 \times 128.9} = 0.1409$$

Note that, in the final stage of dividing the numerator by the denominator, the difference between the characteristics is (0 – 1), which is –1 or $\bar{1}$.

No.	Log
73.56	1.8667
0.095	$\bar{2}$.9777
numerator	0.8444
denominator	1.6954
0.1409	$\bar{1}$.1490
0.3847	$\bar{1}$.5851
128.9	2.1103
denominator	1.6954

Example

0.086 25 ÷ 0.007 21 = 11.96

No.	Log
0.08625	$\bar{2}$.9358
0.00721	$\bar{3}$.8579
11.96	1.0779

Note that the difference between the characteristics is $\bar{2} - \bar{3}$, which may be written as $-2 - (-3) = -2 + 3 = 1$.

6.5 Logarithm of unity

In section 3.3 we saw that a base raised to the power 0 has the value 1, whatever the base.

i.e. $10^0 = 1$

$\therefore \quad \log_{10} 1 = 0$

6.6 Logarithms of reciprocals

Logarithms may be used to calculate the value of reciprocals of numbers, as an alternative to the use of reciprocal tables.

Example Find the reciprocal of 9.16.

$$\text{Reciprocal} = \frac{1}{9.16} = \underline{0.1091}$$

No.	Log
1	0.0000
9.16	0.9619
0.1091	$\bar{1}.0381$ ←

Since log 1 = 0, the subtraction of log 9.16 is a continued process of borrowing 1 from each column until eventually the 1 borrowed from the characteristic column must be written as $\bar{1}$ after subtraction.

Example

$$\frac{1}{0.056} = \underline{17.85}$$

No.	Log
1	0.0000
0.056	$\bar{2}\,7482$
17.85	1.2518 ←

Note that, in the characteristic column, $\bar{2}$ from $\bar{1}$ (1 having been borrowed from the 0) is $-1 - (-2)$ or $-1 + 2 = 1$. (The alternative method of subtraction gives $0 - \bar{1} = 1$.)

6.7 Logarithms of powers

To use logarithms to raise a number to a power, we use the third law of indices: *to raise a power to a power, multiply the indices* (see section 3.2). This is best explained by an example.

Example The square (second power) of 17.54 is required (17.54^2).

log 17.54 = 1.2440

$\therefore$ 17.54 = $10^{1.2440}$ (see section 6.2)

and 17.54^2 = $(10^{1.2440})^2$

Multiply the indices: = 10^{24880}

Antilog: = $\underline{307.6}$

The method is to multiply the logarithm of the number by the index of the power and find the antilogarithm of the result. As with previous calculations with logarithms, the presentation may be simplified by using columns for number and logarithm.

Example

$(1.895 \times 21.43)^2 = \underline{1649}$

No.	Log
1.895	0.2776
21.43	1.3310
	1.6086
× index	2
1649	3.2172

6.8 Logarithms of roots

Roots may be regarded as fractional powers, as we saw in section 3.1. Using logarithms to find a root of a number uses the same law of indices that we used in section 6.7 above. The following example illustrates this.

Example Find the value of $\sqrt{79}$.

$$\log 79 = 1.8976$$

$$\therefore 79 = 10^{1.8976}$$

$$\sqrt{79} = 79^{\frac{1}{2}} = (10^{1.8976})^{\frac{1}{2}}$$

*Multiply the indices: $= 10^{0.9488}$

Antilog: $= \underline{8.888}$

*Multiplication by $\frac{1}{2}$ is equivalent to division by 2, so we may make the rule: *to find the root of a number, divide the logarithm of the number by the index of the root and then find the antilogarithm.*

Example

$\sqrt{87.78} = \underline{9.369}$

No.	Log
87.78	1.9434
÷ 2	2)1.9434
9.369	0.9717

When finding the root of a number less than 1, care is needed when dividing the logarithm by the index of the root because it is not possible to read the negative characteristic and the positive mantissa as a continuous number. The following example illustrates this.

Example

$\sqrt{0.8426} = \underline{0.9179}$

No.	Log
0.8426	$\bar{1}.9256$
÷ 2	2)$\bar{2}$ + 1.9256
0.9179	$\bar{1}$ 0.9628

The division by 2 of the characteristic must be carried out separately from the division of the mantissa. Since an integral result is necessary, the characteristic is changed to 2. To compensate for this, 1 is added to the mantissa. The division of the negative part and the positive part may now be carried out separately and the antilogarithm of the result found.

Note that the root of a number less than 1 is *greater* than the number itself.

6.9 Examples using logarithms

The following problems are typical of problems that may be encountered in the other subjects of the course, and illustrate the use of mathematical tables.

Example The density of a body's material is found by dividing the mass of the body by its volume. Find the density of a piece of steel having a mass of 1.725 kg and a volume of 221 cm^3.
Note. The densities of materials used in construction are usually expressed in kilograms per cubic metre (kg/m^3).

$$\text{Volume} = \frac{221}{10^6}\ m^3$$

(There are 100 cm in 1 m, ∴ $1\ m^3 = 100^3\ cm^3 = 10^6\ cm^3$.)

$$\text{Density} = \frac{\text{mass}}{\text{volume}}$$

$$= \frac{1.725}{\frac{221}{10^6}}\ kg/m^3$$

Invert divisor and multiply: $= 1.725 \times \frac{10^6}{221}\ kg/m^3$

$= 7\,805\ kg/m^3$

No.	Log
1.725	0.2368
10^6	6.0000
	6.2368
221	2.3444
7 805	3.8924

Example The efficiency of a machine is the ratio of 'work got out' to 'work put in'. Calculate the efficiency of a winch that requires an input of 3.84 kilojoules to produce 2.75 kilojoules of useful energy. Express the result as a percentage to the nearest whole number.

$$\text{Efficiency} = \frac{\text{output}}{\text{input}}$$

$$= \frac{2.75}{3.84}$$

$$= 0.7161$$

Multiply by 100% (see 2.6): $=$ 72% (nearest 1%)

No.	Log
2.75	0.4393
3.84	0.5843
0.7161	$\bar{1}.8550$

Example The velocity of a sound is the product of its frequency and its wavelength. Calculate the velocity in metres per second of a sound having a frequency of 2510 hertz and a wavelength of 132 mm.

Note. The unit hertz (Hz) was formerly expressed as cycles per second (c/s). To obtain velocity in m/s, the wavelength must be in metres.

Velocity = frequency × wavelength

= 2510 × 0.131 m/s

= 331.3 m/s

No.	Log
2510	3.3997
0.132	$\bar{1}.1206$
331.3	2.5203

Example The total electrical resistance, R_T, of three resistances R_1, R_2 and R_3 connected in parallel is obtained from the formula:

$$\frac{1}{R_T} = \frac{1}{R_1} + \frac{1}{R_2} + \frac{1}{R_3}$$

What is the total resistance of 4.5 Ω, 6.25 Ω and 5.75 Ω connected in parallel? Note that the symbol Ω (omega) is used for the unit of electrical resistance, the ohm.

$$\frac{1}{R_T} = \frac{1}{4.5} + \frac{1}{6.25} + \frac{1}{5.75}$$

$$= 0.2222 + 0.16 + 0.1739 \quad \text{(reciprocal tables)}$$

$$= 0.5561$$

$$R_T = \frac{1}{0.5561}$$

$$= 1.799\ \Omega \quad \text{(reciprocal tables)}$$

Example The final velocity, v m/s, of a body moving with uniform acceleration, a m/s^2, during a period of time t seconds is given by:

$v = u + at$

where u is its initial velocity in m/s. Find the final velocity after 11.3 seconds of a body moving with uniform acceleration 1.64 m/s^2 from an initial velocity of 2.75 m/s.

$v = u + at$

= 2.75 + (1.64 × 11.3) m/s

= 2.75 + 18.53 m/s

= 21.28 m/s

No.	Log
11.3	1.0531
1.64	0.2148
18.53	1.2679

Example The pressure in kN/m^2 on the subsoil under a foundation slab carrying a uniformly distributed load is found by dividing the total load (force) transmitted to the soil by the area of contact between the slab and the soil. Find the pressure under a rectangular slab 1.65 m × 2.35 m if the total load it transmits to the soil is 54.8 tonnes.

The tonne is a unit of mass equal to 1 000 kg. The force of gravity acting on 1 kg is 9.81 N (newton). It follows that the force of gravity on a mass of 1 tonne, in other words its weight, is 9 810 N.

∴ Weight of 54.8 tonnes $= 54.8 \times 9810$ N

$= 54.8 \times 9.81$ kN

$$\text{Pressure} = \frac{\text{force}}{\text{area}}$$

$$= \frac{54.8 \times 9.81}{1.65 \times 2.35}\ \text{kN/m}^2$$

$$= \underline{138.7\ \text{kN/m}^2}$$

No.	Log
54.8	1.7388
9.81	0.9917
	2.7305
	0.5886
138.7	← 2.1419
1.65	0.2175
2.35	0.3711
	0.5886

Example A brick being tested for its water absorption properties is found to weigh 2.728 kg when dry and 3.135 kg after soaking in water for 24 hours. Calculate its percentage absorption by weight.

Note. The percentage absorption required is the weight of water absorbed expressed as a percentage of the brick's dry weight.

Actual absorption $= 3.135 - 2.728$ kg

$= 0.407$ kg

$$\text{Percentage absorption} = \frac{0.407}{2.728} \times 100\%$$

$= 0.1491 \times 100\%$

$= \underline{15\%}$

No.	Log
0.407	$\bar{1}.6096$
2.728	0.4359
0.1491	← $\bar{1}.1737$

Note the approximation. Brick is a material with varying properties, and to give such a result to any greater degree of accuracy than 1% would be quite unnecessary.

6.10 Limitations of four-figure tables

The accuracy obtained when using four-figure mathematical tables varies with the magnitude of the quantities involved. This is illustrated by the following four calculations involving money. (When absolute precision is required, methods involving greater accuracy will obviously have to be used.)

Example 1 Find the cost of 16 panes of glass at £1.57 each.

Cost = 16 × £1.57

= £25.12

No.	Log
1.57	0.1950
16	1.2041
25.12	1.4000

This involves small numbers and there is no error in the result.

Example 2 Find the cost of converting 23 windows to double glazing at £34.78 per window.

Cost = 23 × £34.78

≏ £799.80 (by logs)

No.	Log
23	1.3617
34.78	1.5413
799.8	2.9030

The correct figure is £799.94, an error of 14p.

Example 3 Calculate the annual income of a man receiving quarterly rents of £114.50 from each of 19 properties.

Income = 19 × £114.50 × 4

≏ £8 706 (by logs)

No.	Log
19	1.2788
114.5	2.0589
4	0.6021
8 706	3.9398

The correct figure is £8 702, an error of £4.

Example 4 What is the total sum received from the sale of 44 houses on a new estate, 23 at £13 750 and 21 at £15 860?

Total sum = £(23 × 13 750) + (21 × 15 860)

≏ £(316 200 + 333 200)

≏ £649 400 (by logs)

No.	Log
23	1.3617
13 750	4.1383
316 200	5.5000
21	1.3222
15 860	4.2004
333 200	5.5226

The correct sum is £649 310, an error of £90.

Exercise 6c Find the numbers whose logarithms are:

1. 1.8436 2. 2.0193 3. 0.7561

4. $\bar{1}.5241$ 5. $\bar{2}.1817$ 6. 3.6225

Evaluate the following using four-figure logarithms:

7. 36.84×1.732

8. $9.81 \times 3.142 \times 17.85$

9. $85.93 \div 2.65$

10. $254.5 \div 32.75$

11. $\dfrac{47.63 \times 3.81}{15.24 \times 2.76}$

12. $\dfrac{0.079 \times 586.4}{0.003\,95 \times 293.2}$

13. $\dfrac{1}{53.76 \times 0.726}$

14. $\dfrac{1}{3489}$

15. 7.35^2 16. 0.524^2 17. 27.59^2

18. 3.562^3 19. 2.518^4 20. 0.0765^2

21. $\sqrt{593}$ 22. $\sqrt{11.95}$ 23. $\sqrt{0.6321}$

24. $\sqrt{0.009\,54}$ 25. $\dfrac{\sqrt{23} \times 1.54^2}{1.101^3}$ 26. $\dfrac{16.13^2 \times 3.142}{\sqrt{6.18}}$

27. Pressure p in newtons per square metre at a depth h metres in a liquid of specific weight w newtons per cubic metre is the product of h and w. Find the pressure at a depth of 1.523 m in a liquid of specific weight 9 810 N/m^3.

28. A straight-flight staircase has a total rise of 2.250 m and a total going of 2.580 m. There are 13 steps. Calculate the rise and the going of each step in mm to the nearest mm.

29. A window is 1 350 mm high and 1 050 mm wide. What must be the height of a second window 985 mm wide if it is to be in the same proportion? Give your answer to the nearest mm.

30. The internal diameter of a tube is 37.5 mm and its wall thickness is 2.4 mm. What must be the wall thickness of a tube with an internal diameter of 43.5 mm if they are to have the same proportions?

7 THE SLIDE RULE

7.1 The construction of a slide rule

A *slide rule* (Figure 1) is a mechanical means of *adding and subtracting logarithms* and therefore of *multiplying and dividing numbers.*

The scales are *graduated in logarithms* of numbers but *marked with the numbers.* A logarithmic scale is illustrated in Figure 2.

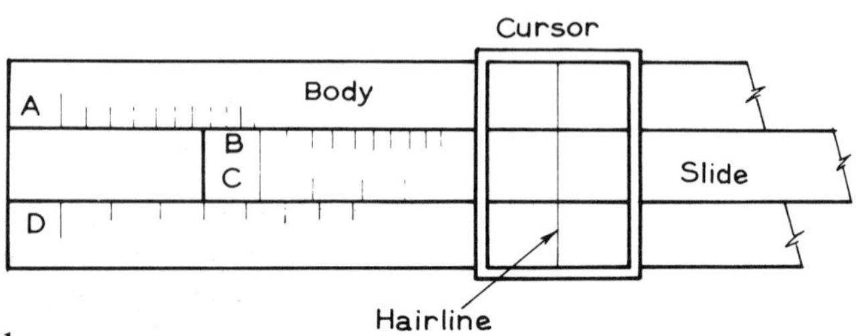

Figure 1

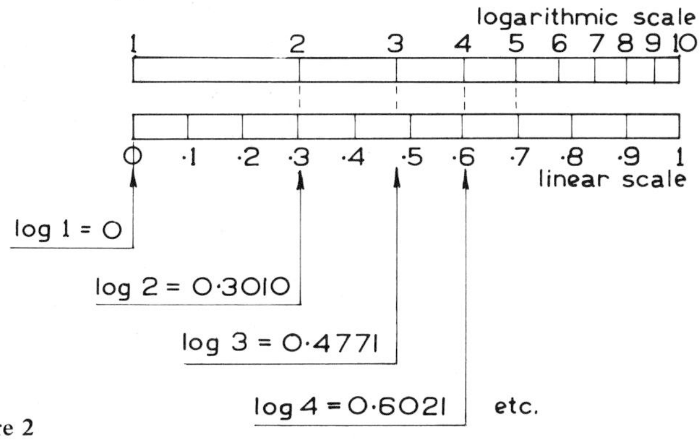

Figure 2

Scale A on the upper side of the body and scale B on the top edge of the slide are identically graduated in logarithms of numbers from 1 to 100.

Scale C on the lower edge of the slide and scale D on the lower side of the body are graduated in logarithms of numbers from 1 to 10. The C and D scales are therefore double the A and B scales, thus allowing greater accuracy.

7.2 Multiplication and division

MULTIPLICATION

1.7 x 2.1 (see Figure 3)

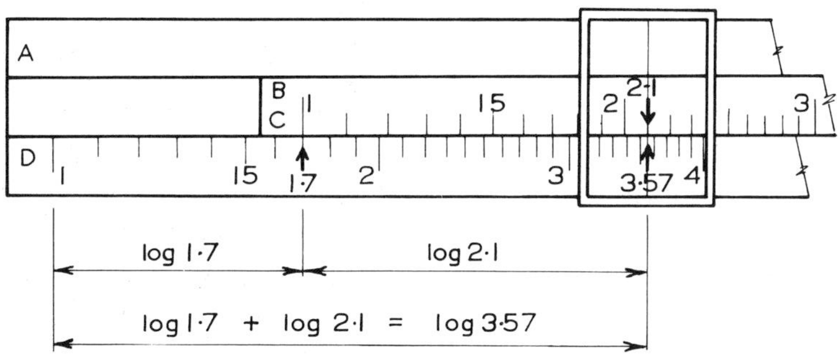

Figure 3

Although it is the *logarithms* that are added, it is the *numbers* that are read.

e.g. log 1.7 + log 2.1 = log 3.57

∴ 1.7 x 2.1 = 3.57

Continued multiplication is merely a series of repetitions of the above process, each product obtained being multiplied by the next multiplier.

Example

1.2 x 1.9 x 2.6

Place the 1 index of scale C at 1.2 on scale D. Move the cursor until the hairline is on 1.9 on scale C. The product of 1.2 x 1.9 is now indicated by the hairline on scale D.

Move the slide until the 1 index of scale C coincides with the hairline. Move the cursor until the hairline is at 2.6 on scale C. The product of 1.2 x 1.9 x 2.6 is now indicated by the hairline on scale D. It is 5.93.

∴ 1.2 x 1.9 x 2.6 = 5.93

A situation will be met where the required operation would take the cursor off the end of the body. When this occurs the slide should be used from its right-hand end; i.e. place the 10 index on the C scale against the multiplicand on the D scale, and then move the cursor until the hairline coincides with the multiplier as before. The product is then read off from the hairline on the D scale.

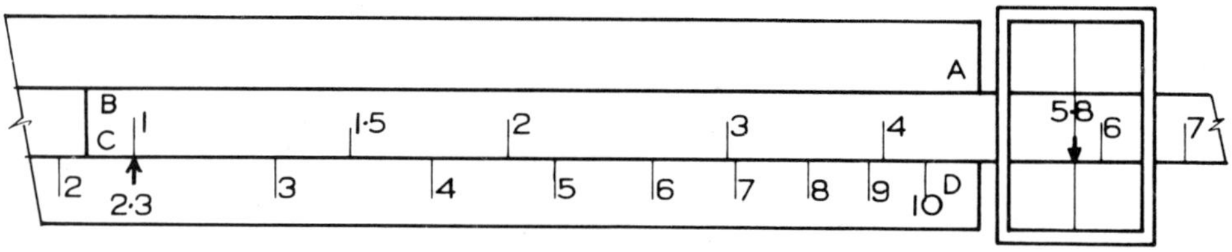

Left index of C scale used — cursor off end

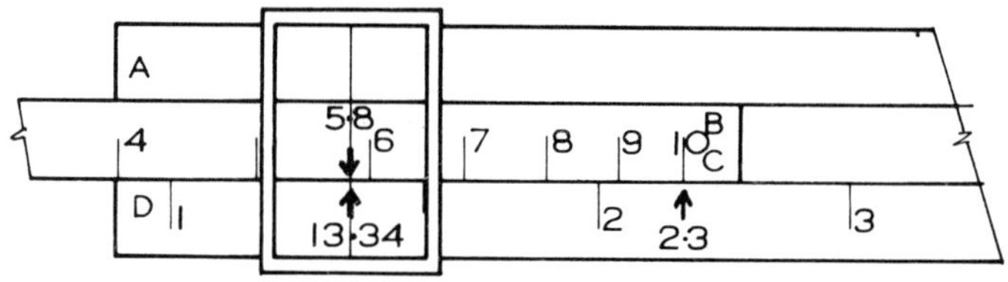

Figure 4 Right index of C scale used

Example

2.3 x 5.8 (see Figure 4)

The slide rule does not indicate the position of the decimal point. This may be obtained by an approximate estimation of the result of the calculation (see section 5.3).

If we approximate 2.3 to 2 and 5.8 to 6, we obtain an approximate product of 12. The result from the D scale is therefore 13.34.

∴ 2.3 x 5.8 = 13.34

The decision whether to use the left index (1) or the right index (10) of the C scale when multiplying may be made as follows: *If the approximate product is less than 10 use the left index, if greater than 10 use the right index.*

Now try questions 1 to 12 of Exercise 7a (page 49).

DIVISION

The logarithm of the divisor is subtracted from the logarithm of the dividend to obtain the logarithm of the quotient.

8.6 ÷ 3.5 (see Figure 5)

The rough check 8 ÷ 3 gives 2.7. The slide rule gives the figures 246.

∴ 8 ÷ 3 = 2.46

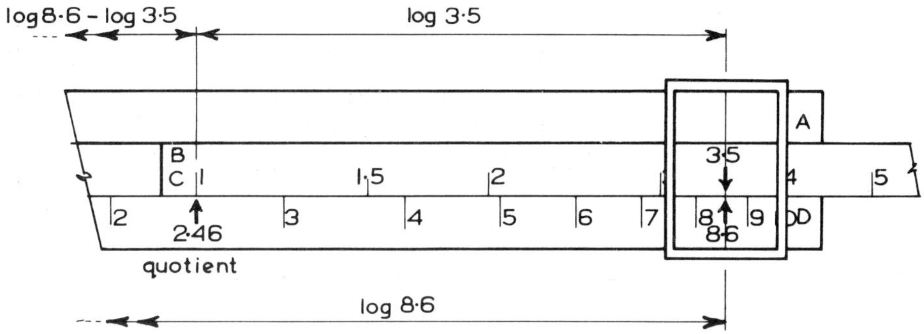

Figure 5

Division by more than one divisor is a continuous process of similar operations.

Example

$$\frac{84.5}{21.3 \times 3.75}$$

First a rough estimate:

$$\frac{84}{21 \times 4} = 1$$

1st operation (Figure 6): set the hairline to 845 on the D scale, and move the slide until 213 on the C scale coincides. This gives the first quotient at the index of the C scale (its value is not required).

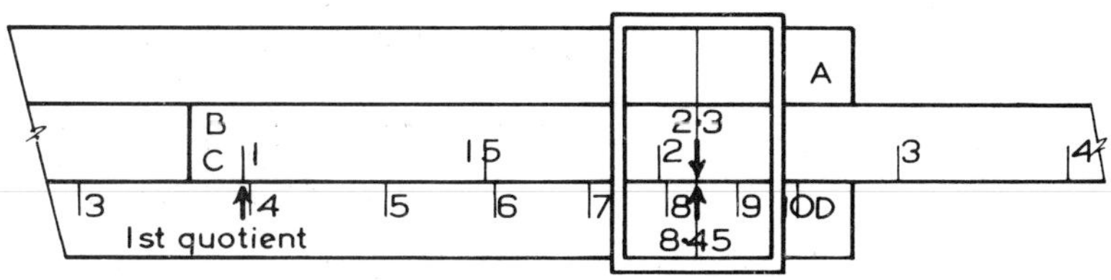

Figure 6

2nd operation (Figure 7): move the cursor until the hairline coincides with the first quotient, and then move the slide until 375 on the C scale is under the hairline. The second quotient (the result) is now read at the index of the C scale. This is 106.

$$\therefore \quad \frac{84.5}{21.3 \times 3.75} = 1.06$$

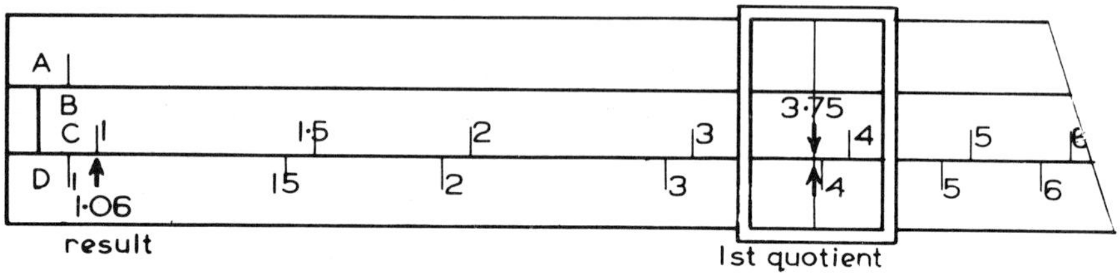

Figure 7

Now try questions 13 to 18 of Exercise 7a (page 49).

COMBINED MULTIPLICATION AND DIVISION

An expression containing mixed operations of multiplication and division is best evaluated by a 'zig-zag' method as follows.

Example

$$\frac{3.5 \times 58.6}{21.4 \times 12.75}$$

First an estimate:

$$\frac{3 \times 60}{20 \times 12} = \frac{3}{4} = 0.75$$

1st operation (Figure 8): divide 3.5 by 21.4 by setting the hairline to 35 on the D scale and moving the slide until 214 on the C scale coincides. This gives the first quotient on the D scale (its value is not required).

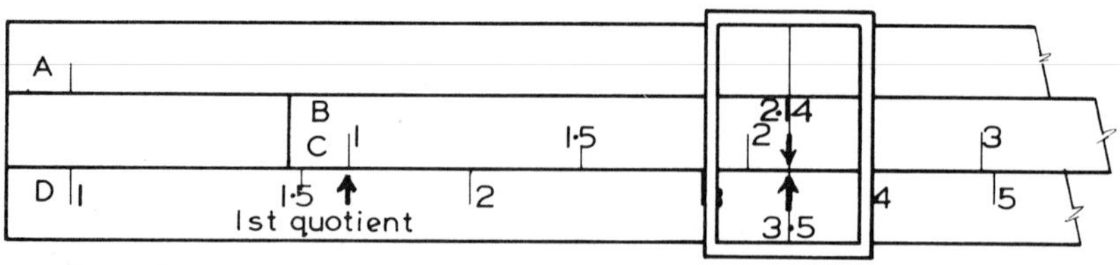

Figure 8

2nd operation (Figure 9): multiply the first quotient by 58.6 by moving the cursor until the hairline coincides with 586 on the C scale. This gives the first product on the D scale (its value is not required).

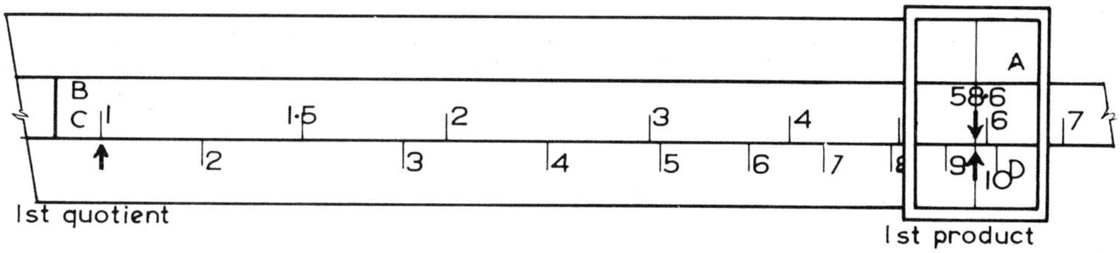

Figure 9

3rd operation (Figure 10): divide the first product by 12.75 by moving the slide until 1275 on the C scale coincides with the hair-line. Read off the second quotient (the result) at the index of the C scale. The slide rule reading is 752.

$$\therefore \quad \frac{3.5 \times 58.6}{21.4 \times 12.75} = 0.752$$

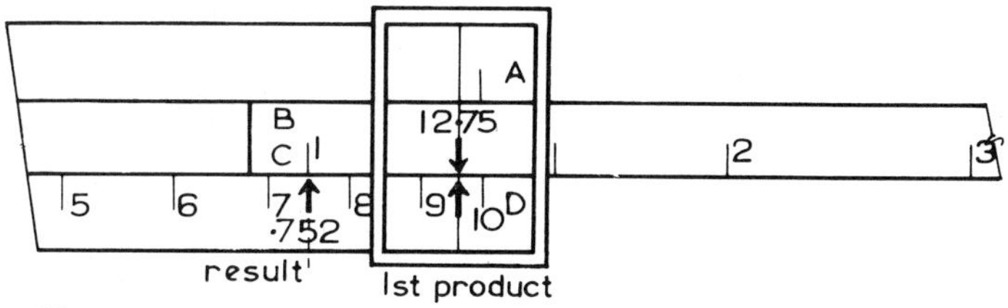

Figure 10

Now try questions 19 to 23 in the following exercise.

Exercise 7a Using a slide rule, evaluate the following products:

1. 184 × 2.83 2. 59.3 × 275 3. 1.56 × 856
4. 0.78 × 4.15 5. 0.053 × 0.925 6. 1325 × 0.0082
7. 8.2 × 3.4 × 5.5 8. 0.024 × 6.83 × 13.7
9. 4.6 × 52.7 × 11.2 10. 0.631 × 11.8 × 9.75 × 1.28
11. 0.005 5 × 0.032 5 × 6 × 12 100 × 3 100
12. 10.56 × 0.067 × 0.75 × 1 320 × 0.54

Find the values of the following quotients:

13. $\frac{580}{14}$ 14. $\frac{3\,760}{23.6}$ 15. $\frac{95.8}{32.9}$

16. $\frac{0.062\,1}{0.008\,4}$ 17. $\frac{176}{7\,320}$ 18. $\frac{1\,420}{896}$

19. $\frac{17 \times 375}{56}$ 20. $\frac{27.5 \times 38.2}{1\,820}$ 21. $\frac{4.6 \times 362}{412 \times 2.76}$

22. $\frac{0.063 \times 5.9 \times 11.4}{32.2 \times 7.56}$ 23. $\frac{8\,700}{386 \times 1.09 \times 65}$

7.3 Squares, square roots and reciprocals

SQUARES

Since the A scale is graduated from 1 to 100 and the D scale from 1 to 10, it is possible to find squares of numbers marked on the D scale by reading directly across to the A scale. The position of the decimal point must be determined independently.

Example

37.4^2

Set the hairline to 374 on the D scale. Read off the coinciding number on the A scale. This is 14.

Since 37.4 is between 30 and 40, its square must lie between 900 and 1 600.

$\therefore\ \underline{37.4^2 = 1400}$

Example

$$\left(\frac{48.5 \times 8.12}{516}\right)^2$$

Approximation $\left(\frac{50 \times 8}{500}\right)^2 = (0.8)^2 = 0.64$

1st operation: divide 48.5 by 516 by setting the hairline to 485 on the D scale and moving the slide until 516 coincides. The 1st quotient is indicated at the 10 index of scale C on scale D (its value is not required).

2nd operation: multiply the 1st quotient by 8.12 by moving the cursor until the hairline coincides with 812 on the C scale. This gives the product on the D scale.

3rd operation: without any adjustment of the slide rule, read off the square of the product on the A scale. The figures are 583.

$$\therefore\ \underline{\left(\frac{48.5 \times 8.12}{516}\right)^2 = 0.583}$$

SQUARE ROOTS

The reverse process, reading direct from the A scale to the D scale, will give square roots. Care must be taken over this operation, for the reasons given in section 6.1.

For numbers with an odd number of digits in their integral part, use the A scale from 1 to 10 (the left half). For numbers with an even number of digits, use the right half (10 to 100).

Example

$\sqrt{58.5}$

Place the cursor with the hairline on 585 on the right half of scale A. Read off the square root on scale D. This is 765, but the decimal point now has to be positioned.

Since there is one pair of digits in the number, there will be one digit in its square root.

$\therefore$ $\underline{\sqrt{58.4} = 7.65}$

Example

$\sqrt{87\,600}$

Place the cursor with the hairline on 876 on the left half of scale A and read off the square root on the D scale. The figures are 296. There are two pairs of digits and one 'odd' digit in the number, so there will be three digits in the square root.

$\therefore$ $\sqrt{87\,600} = 296$

RECIPROCALS

To obtain reciprocals of numbers we may *either* divide the number into 1 using the C and D scales *or* read the reciprocal directly from the C1 scale.

For the first method, simply move the slide until the number coincides with either index, the 1 or the 10, and read off the reciprocal at the opposite end index.

The second method is simpler. The C1 scale is identical to the C scale but printed or engraved from right to left. Numbers on the C scale therefore coincide with their reciprocals on the C1 scale. The only difficulty is familiarising oneself with reading from right to left.

Exercise 7b Find the following squares and square roots:

1. 11.5^2	2. 0.65^2	3. 0.076^2
4. 5.15^2	5. 327^2	6. 41.5^2
7. $\sqrt{8.13}$	8. $\sqrt{172}$	9. $\sqrt{91.8}$
10. $\sqrt{0.068}$	11. $\sqrt{0.005\,3}$	12. $\sqrt{9\,350}$

Find the values of the reciprocals:

13. $\dfrac{1}{0.075}$	13. $\dfrac{1}{1.65}$	15. $\dfrac{1}{12.8}$
16. 6.75^{-1}	17. 125^{-1}	18. 82.5^{-1}

7.4 Evaluating compound expressions

Only by practice will speed of operation and accuracy be achieved. The following examples illustrate mixed operations in slide rule calculations, and you are advised to follow them on your own slide rule. To assist you in this, the intermediate evaluations that are not normally read off are shown in brackets at each stage. The decimal point is ignored after the approximation until completion.

Example

$$\frac{365 \times \sqrt{3.75} \times 4}{11.6 \times 5.3}$$

Approximation $\dfrac{360 \times 2 \times 4}{12 \times 5} = 48$

1st operation: since $\sqrt{3.75}$ will be obtained by reading from the A scale to the D scale, do this first. The remaining operations may then be carried out on the C and D scales.

Set the hairline to 375 on the left half of the A scale. The square root is opposite on the D scale (194).

2nd operation: move the slide until 116 on the C scale coincides with the hairline (quotient 167).

3rd operation: move the cursor until the hairline coincides with 365 on the C scale (product 61).

4th operation: move the slide until the 5.3 on the C scale coincides with the hairline (quotient 115).

5th operation: move the cursor until the hairline coincides with the 4 on the C scale. Read off the result on the D scale (46).

$$\therefore \quad \frac{365 \times \sqrt{3.75} \times 4}{11.6 \times 5.3} = 46$$

Example

$$\left(\frac{56 \times 38 \times \sqrt{1.75}}{\sqrt{2} \times 76.6}\right)^2$$

This expression lends itself to simplification before we begin to use the slide rule.

$(\sqrt{1.75})^2 = 1.75$

$(\sqrt{2})^2 = 2$

The expression may therefore be rewritten as:

$$\frac{1.75}{2}\left(\frac{56 \times 38}{76.6}\right)^2$$

Approximation $\quad 0.8 \times \left(\dfrac{60 \times 40}{80}\right)^2$

$= 0.8 \times (30)^2$

$= 0.8 \times 900 = 720$

1st operation: set the hairline to 56 on the D scale.

2nd operation: move the slide until 766 on the C scale coincides (quotient 731).

3rd operation: move the cursor until the hairline coincides with 38 on the C scale (product 278).

4th operation: transfer to the A scale for the square of this product (772).

5th operation: move the slide until the 100 index on the B scale coincides with the hairline.

6th operation: move the cursor until the hairline coincides with the 875 on the B scale. Read off the resulting product on the A scale (675).

$$\therefore \left(\frac{56 \times 38 \times \sqrt{1.75}}{\sqrt{2} \times 76.6}\right)^2 = 675$$

Exercise 7c Evaluate the following expressions:

1. $\dfrac{5.76^2 \times 2.25 \times 53}{763 \times 1.89}$
2. $\dfrac{110 \times \sqrt{5.4} \times 27^2}{1400 \times 10.5}$
3. $\dfrac{\sqrt{52} \times 3.21^2}{\sqrt{8.9}}$
4. $\dfrac{\sqrt{82} \times 3.54}{11.3}$

8 CALCULATING MACHINES

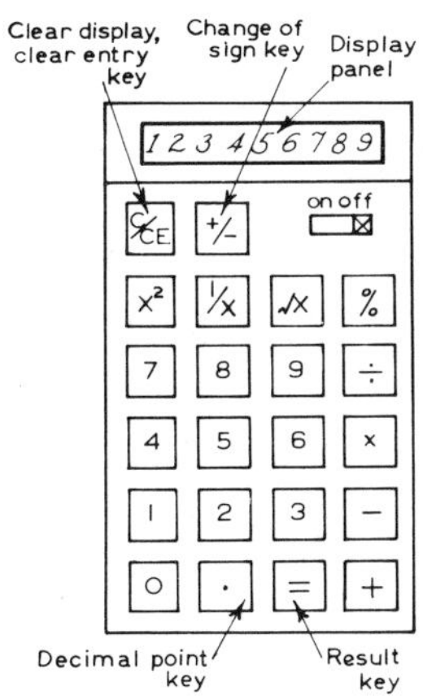

Figure 11

Electronic calculators have been developed to high standards, and the simpler versions are obtainable at comparatively low prices. Evaluations by calculator have the advantages of ease of manipulation, speed of working and high degree of accuracy. This assumes that the operator is accurate in his use of the instrument and that the battery is not discharged.

A calculator is particularly useful when a large number of repetitive operations have to be made. Models vary from very simple ones capable of addition and subtraction, multiplication and division, to extremely complex ones for scientific applications.

The diagram (Figure 11) does not represent any particular model but typifies some of the popular models, and the examples that follow make reference to this diagram. The majority of the keys are explained by their labels, but brief notes have been added to describe others. Some of the simpler models do not have the x^2, $\sqrt{x}$ and $1/x$ keys.

Manufacturers provide instruction manuals, which should be studied carefully to obtain the best results.

8.1 Basic operations

ADDITION AND SUBTRACTION

The sequence of operations for addition and subtraction follows the normal procedure as it would be written on paper. This is termed *algebraic logic*.

Example

7 + 4

Sequence of operations:

[7] [+] [4] [=]

The display panel shows each new number entered as the keys are pressed. The result key, pressed at the end of the sequence, produces the result on the panel. The above operation would appear as follows.

Key pressed		*Display*
[7]		7
[+]		7
[4]		4
[=]	Result:	11

The [C/CE] key clears the calculator for the next operation.

Example

6 + 9 − 3

Sequence of operations:

[6] [+] [9] [−] [3] [=]

After the first operation (6 + 9), the pressing of the next function key ([−]) displays the result of the previous calculation (15).

Key pressed		*Display*
[6]		6
[+]		6
[9]		9
[−]		15
[3]		3
[=]	Result:	12

Errors made in entering digits may be corrected immediately by pressing the [C/CE] key, which clears the entry.

Example

5 − 6 + 14 − 7

Correct sequence of operations:

[5] [−] [6] [+] [1] [4] [−] [7] [=]

Suppose that an error is made by entering a [4] instead of the [1]. The operation, including the correction, would then appear as follows:

Key pressed		*Display*
[5]		5
[–]		5
[6]		6
[+]		–1
[4]	Error made:	4
[C/CE]	Clear entry:	0
[1]	Correction:	1
[4]		14
[–]		13
[7]		7
[=]	Result:	6

MULTIPLICATION AND DIVISION

These operations also follow algebraic logic, the numbers being entered in the sequence in which they are written.

Example

15 x 7

Sequence of operations:

[1] [5] [×] [7] [=]

Key pressed		*Display*
[1]		1
[5]		15
[×]		15
[7]		7
[=]	Result:	105

As with addition and subtraction, the pressing of the second and subsequent function keys displays the previous calculation made.

Example

5.5 x 2.63 x 12.6

Sequence of operations:

[5] [.] [5] [×] [2] [.] [6] [3] [×] [1] [2] [.] [6] [=]

Key pressed		*Display*
[5]		5
[.]		5
[5]		5.5
[×]		5.5
[2]		2
[.]		2
[6]		2.6
[3]		2.63
[×]	Previous calculation:	14.465
[1]		1
[2]		12
[.]		12
[6]		12.6
[=]	Result:	182.259

Example

516 ÷ 28.5

Sequence of operations:

[5] [1] [6] [÷] [2] [8] [.] [5] [=]

Key pressed	*Display*
[5]	5
[1]	51

[6]		*516*
[÷]		*516*
[2]		*2*
[8]		*28*
[.]		*28*
[5]		*28.5*
[=]	Result:	*18.105263*

This result contains eight digits and six decimal places. Such accuracy is unlikely to be required under normal circumstances and correction to the required number of significant figures or decimal places should be made, say 18.11.

Combined multiplication and division operations may be made in any order, i.e. in zig-zag order as on a slide rule or in continuous order as read.

Example

$$\frac{17 \times 5.6}{2.8}$$

Sequence of operations:

[1] [7] [×] [5] [.] [6] [÷] [2] [.] [8] [=]

Key pressed		*Display*
[1]		*1*
[7]		*17*
[×]		*17*
[5]		*5*
[.]		*5*
[6]		*5.6*
[÷]	Previous calculation:	*95.2*
[2]		*2*
[.]		*2*

[8]		2.8
[=]	Result:	34

When the denominator contains more than one number, it must be remembered that each is a divisor and entry must be preceded by the ÷ sign.

Example

$$\frac{125.6}{4 \times 15.7}$$

Sequence of operations:

[1] [2] [5] [.] [6] [÷] [4] [÷] [1] [5] [.] [7] [=]

Both the 4 and the 15.7 are divisors.

Key pressed		*Display*
[1]		1
[2]		12
[5]		125
[.]		125
[6]		125.6
[÷]		125.6
[4]		4
[÷]	Previous calculation:	31.4
[1]		1
[5]		15
[.]		15
[7]		15.7
[=]	Result:	2

8.2 Powers Powers of numbers may be found by continued multiplication, but this can be short cut as follows.

Example

$(3.5)^4$

Sequence of operations:

[3] [.] [5] [×] [=] [=] [=]

The number is entered, followed by the × sign, and the result key is pressed one time less than the index of the power required. (Some calculators have a 'constant' key that must be pressed prior to this sequence.)

Key pressed		*Display*
[3]		3
[.]		3
[5]		3.5
[×]		3.5
[=]	$(3.5)^2$	12.25
[=]	$(3.5)^3$	42.875
[=]	$(3.5)^4$	Result: 150.0625

Many calculators have an [x^2] key enabling the second power of a number to be found directly. The number is entered and the [x^2] key pressed to display its square.

8.3 Reciprocals Reciprocals of numbers may be found directly if the calculator has a [$1/x$] key. Alternatively, a division sequence is used with 1 as the dividend and the number as the divisor.

Example

$$\frac{1}{25}$$

Sequence of operations:

[1] [÷] [2] [5] [=]

Key pressed	*Display*
[1]	1
[÷]	1
[2]	2

[5] 25

[=] Result: 0.04

If the calculator has a reciprocal key, it is pressed after entering the number. The reciprocal of the number is then displayed.

8.4 Practical problems

Calculations arising from problems in science and technology are readily carried out on he calculator. It is important to marshal the facts carefully and arrange the correct expressions before the calculator is used. The calculator cannot make mistakes: only the operator can do this!

Examples of typical calculations arising in Unit 1 work are given below.

Example A length of flat steel bar 300 mm long has cross-sectional dimensions of 52.5 mm × 12.5 mm and a mass of 1.54 kg. Calculate its relative density correct to two decimal places.
Note. Relative density is the material's density compared to the density of water. Since the density of water is 1 g/cm^3, a convenient approach is to calculate the density in g/cm^3; the relative density then has the same numerical value but, of course, without units.

Mass in grams = 1540 g

Volume in cm^3 = $30 \times 5.25 \times 1.25$ cm^3

$$\text{Density} = \frac{1540}{30 \times 5.25 \times 1.25} \text{ g/cm}^3$$

Relative density = 7.82

Sequence of operations for the above calculation:

[1] [5] [4] [0] [÷] [3] [0] [÷] [5] [.] [2] [5] [÷] [1] [.] [2] [5] [=]

Result: 7.82

Example The tensile stress in a steel tie-bar is not to exceed 150 N/mm^2. What is the maximum safe load the bar may carry if its cross-sectional dimensions are 112 mm × 15 mm?

Load = stress × area

$$= 150 \times 112 \times 15 \text{ N}$$

$$= \frac{150 \times 112 \times 15}{1\,000} \text{ kN}$$

$$= 15^2 \times 1.12 \text{ kN}$$

Sequence of operations:

[1] [5] [x^2] [×] [1] [.] [1] [2] [=]

Result: 252 kN

Example A pump is capable of raising water to a height of 8 m at the rate of 550 litres per minute. Calculate the power developed.

$$\text{Power} = \frac{\text{work done}}{\text{time taken}}$$

$$= \frac{\text{force} \times \text{distance}}{\text{time}}$$

$$= \frac{550 \times 9.81 \times 8}{60} \text{ N m/s}$$

Note. 1 litre of water has a mass of 1 kg.
Weight of a 1 kg mass = 9.81 N
1 N m/s = 1 J/s = 1 W

Sequence of operations:

[5] [5] [0] [×] [9] [.] [8] [1] [×] [8] [÷] [6] [0] [=]

Result: 719.4 W

8.5 Comparison of calculation methods

Complete dependence on aids to calculation is unwise. It is often possible to make accurate evaluations mentally, and this should be done whenever possible. The human brain is capable of much more than many people give it credit for, but it will not function at its best without constant application and effort.

Electronic calculators provide a means of quick, accurate results when properly serviced and used, but battery condition is critical.

Slide rules give quick results once their manipulation is mastered, and their degree of accuracy (though not high) is adequate for many applications.

Four-figure mathematical tables are quick and easy to use, readily available and cheap. For most purposes required by the technician their degree of accuracy is adequate.

Exercise 8

(Use an electronic calculator.) Evaluate:

1. $18.75 + 34.85 - 9.837 + 15.36 - 24.84$
2. £327.50 + £12.95 + £54.63 + £17.10 + £154.42
3. $(0.759 + 1.654 + 3.6) - (1.25 + 2.175 + 0.625)$
4. $(63.54 - 61.76 + 14.35)^2$
5. $7.25^2 + 18.93 \quad 57.15$

Find the values of the following expressions correct to six significant figures:

6. 3.142×6.59
7. 118.73×0.625
8. $53.4 \times 1.732 \times 0.41$
9. $0.51 \times 0.86 \times 1.532$
10. $76.54 \div 14.5$
11. $3265.5 \div 162.85$
12. $\dfrac{83.76 \times 14.93}{386.9}$
13. $\dfrac{5.32^2 \times 18.56}{11.72}$

Find the following powers:

14. 1.875^2 15. 0.86^2 16. 2.35^3

17. 127^2 18. 0.975^3 19. 2.07^4

Evaluate the reciprocals:

20. $\frac{1}{17}$ 21. $\frac{1}{23.5}$ 22. $\frac{1}{327}$

23. The safe load for a tie-bar is given by the product of its cross-sectional area and the safe working stress for the material. Calculate the safe load for a tie-bar having a cross-section 75 mm x 12.5 mm and a safe working stress of 165 N/mm^2.

24. Find the density in kg/m^3 of the material of a cube having an edge length of 150 mm and a mass of 8.375 kg.

25. What power is developed when a mass of 250 kg is lifted through a height of 22.5 m in 35 seconds?

26. How much heat is given up by 50 litres of water as it cools from 90 °C to 35 °C? (Specific heat capacity = 4 180 J/kg °C).

SUMMARY OF B

Make rough checks of expected results.
Limit results to a practical degree of accuracy.
Always check the validity of a result.
Reciprocals: use reciprocal tables with care: ***subtract*** the mean differences.
Logarithms: common logarithms are to the base 10.
The common logarithm of a number is the index of the power to which 10 must be raised to give the number.
To multiply, add logarithms.
To divide, subtract logarithms.
To raise a number to a power, multiply the logarithm of the number by the index of the power.
To find a root of a number, divide the logarithm of the number by the index of the root.

REVISION EXERCISE B

Suggest a suitable degree of accuracy for the results of the following expressions:

1. 5.3×8.5 2. $\frac{8.625 \times 3.14}{4.76}$ 3. $\frac{5.94 \times 16}{15.03}$

Obtain approximate values of the following:

4. $\frac{247.6 \times 52.35}{26.45 \times 8.91}$ 5. $\frac{32.46 \times 10.85}{93.4 \times 2.2}$

Use the standard-form method to obtain approximate evaluations of the following:

6. $\frac{1.195 \times 92.8 \times 103.7}{524}$ 7. $\frac{518 \times 13\,255}{83.62 \times 315}$

Use four-figure logarithms to evaluate:

8. $\dfrac{52.63 \times 0.09^2}{\sqrt{0.981}}$ 9. $\dfrac{3\,725 \times 0.0023^2}{\sqrt{1.111}}$

Use logarithms to complete the calculations for problems 10 to 12. (As an additional exercise the results obtained may be compared with the results of using an electronic calculator.)

10. Ohm's law connects the current I in amps, the potential difference V in volts and the resistance R in ohms in a circuit, and is expressed in the formula:

$V/I = R$

Find the resistance of an appliance carrying a current of 8.62 amps at a potential difference of 238 volts.

11. A body dropped freely from a height falls with an acceleration due to gravity of 9.81 m/s^2 and the distance, h metres, through which it falls in a time of t seconds is given by:

$h = \frac{1}{2}gt^2$

Find the distance through which a body will fall in 5.45 seconds.

12. Water absorption tests on slate samples required a result of not more than 0.5% by weight after twelve hours soaking. Calculate the results for the following three specimens:

Specimen no.	1	2	3
Dry weight	378.5 g	381 g	374.5 g
Wet weight	382 g	382.2 g	376.3 g

Use an electronic calculating machine to complete the following four problems:

13. If electrical power is given by the square of the current multiplied by the resistance, and electrical energy is the product of power and time, calculate the quantity of heat generated when a current of 6 amps flows through a resistance of 10 ohms for a period of 4 minutes.

14. The distance h metres through which a body will fall freely from rest in a time of t seconds is given by:

$h = \frac{1}{2}gt^2$

where g is the constant acceleration due to gravity (= 9.81 m/s^2). At what height should a body be released if it is to reach the ground in 10 seconds?

15. What is the pressure at the base of a tank containing liquid having a relative density of 0.85 to a depth of 3.75 m?

16. Calculate the quantity of heat required to convert 10 kg of ice at -18 °C to water at boiling point. The specific heat capacity of ice is half that of water, which is 4.18×10^3 J/kg °C. The specific latent heat of fusion of ice is 3.36×10^5 J/kg. Give the result in MJ to the nearest kJ.

C Algebra

This chapter deals with:
Symbolic representation
Basic algebraic operations; commutative, associative and distributive laws
Laws of indices
Precedence rules
Factorisation of algebraic expressions
Simple equations, simultaneous equations
Evaluation and transformation of formulae

9 ALGEBRAIC NOTATION AND METHODS

9.1 Symbolic representation

The use of symbols in place of numbers allows *general* mathematical statements to be made in place of *specific* ones. Letters are normally used for this purpose, and many letters of the Greek alphabet have been adopted as standard symbols. Probably the best known example is π (pi), which represents the ratio between the circumference and diameter of a circle. Others are θ (theta), ϕ (phi), α (alpha) and β (beta); Ω (omega) is the symbol for the ohm, the unit of electrical resistance. There are many other similar examples of *symbolic representation.*

Suppose a firm producing precast concrete components decides to charge £8.50 for a certain type of slab and £12.75 for a prestressed lintel. Denoting the number of slabs sold as x and the number of lintels as y, the sales income from these two products may be represented as:

£$(8.5x + 12.75y)$

This is an *algebraic expression*, the value of which may be determined only when the values of x and y are known. The expression contains two *terms*, $8.5x$ and $12.75y$. Expressions containing two terms are *binomial* expressions.

If a building site employs 8 craftsmen each earning £a per week, 5 labourers each earning £b per week and one supervisor earning £c per week, the total weekly wage bill for the site may be expressed as:

£$(8a + 5b + c)$

This expression contains three terms and is a *trinomial* expression. Expressions with more than three terms are *polynomial* expressions.

9.2 Manipulation of algebraic expressions

The manipulation of algebraic terms and expressions must obey the same basic laws as the manipulation of numbers.

ADDITION AND SUBTRACTION

$a + b$ represents the sum of two quantities

$a - b$ represents the difference between two quantities

The *associative law* applies, i.e. 'the terms of an expression may be grouped in any order'.

e.g. $(a + b) + c = a + (b + c)$

The *commutative law* applies, i.e. 'additions and subtractions may be made in any order'.

e.g. $a + b = b + a$

Like terms may be collected and combined to a single term.

e.g. $2a + 8a - 4a = 6a$

Unlike terms cannot be combined but must be grouped with + or – signs between them.

e.g. $2a + 3b - a + 7b = a + 10b$

The result of combining a number of like terms in this way, having regard to their *sign*, is called their *algebraic sum.*

Example Find the algebraic sum of the following terms: a, $3b$, $2c$, $3ab$, $4bc$, $7a$, $-2b$, $-2ab$, $-3bc$ and $-c$.

Algeb. sum $= a + 7a + 3ab - 2ab + 3b - 2b + 4bc - 3bc + 2c - c$

$= \underline{8a + ab + b + bc + c}$

Example Subtract $6a + 3b - 4c$ from $8a + 2b + c$.

The expressions may be arranged in vertical order and subtracted.

$$\begin{array}{l} 8a + 2b + c \\ \underline{6a + 3b - 4c} \\ \underline{2a - b + 5c} \end{array}$$

Alternatively, the expressions may be arranged in line as in the first example.

$$\begin{aligned} & 8a + 2b + c - (6a + 3b - 4c) \\ = & 8a + 2b + c - 6a - 3b + 4c \\ = & \underline{2a - b + 5c} \end{aligned}$$

The brackets were inserted round the second expression to indicate that the whole expression was to be subtracted. The 'minus' sign before the bracket changed the sign of each term within the bracket.

MULTIPLICATION AND DIVISION

The multiplication sign is rarely used in algebraic expressions. The product of a and b is usually written as ab, not as $a \times b$.

The *associative law* applies, i.e. 'the factors of a product may be grouped in any way'.

e.g. $a(bc) = (ab)c$

The *distributive law* applies, i.e. 'the product of a compound expression and a single factor is the algebraic sum of the products of the single factor and the terms of the expression'.

e.g. $a(b + c) = ab + ac$

The *commutative law* applies, i.e. 'the factors of a product may be taken in any order'.

e.g. $ab = ba$

The sign ÷ for division is rarely used; instead the divisor is expressed as the denominator of a fraction or ratio,

i.e. $\dfrac{a}{b}$ or a/b instead of $a \div b$

The rules of signs specified for the manipulation of numbers apply equally to algebraic manipulation; i.e. like signs give + results, unlike signs give − results.

e.g.
$$5x(6xy^2) = 30x^2y^2$$
$$7x(3x - 2y) = 21x^2 - 14xy$$
$$-abc(-3ac^2) = 3a^2bc^3$$

The word *coefficient* means *a multiplier*, and each number or symbol in a term is the coefficient of the remainder of the term. In the term $4x^2y$, for example, 4 is the coefficient of x^2y, x^2 is the coefficient of $4y$, and y is the coefficient of $4x^2$. The 4 is a *numerical coefficient.*

The laws of indices previously expressed in numerical terms may now be restated in general terms. (See section 3.2.)

First law of indices: *to multiply, add the indices.*

$$a^m . a^n = a^{m+n}$$

Note. The use of a full stop between the terms is often a convenient alternative to the multiplication sign.

Second law of indices: *to divide, subtract the indices.*

$$a^m / a^n = a^{m-n}$$

Third law of indices: *to raise a power to a power, multiply the indices.*

$$(a^m)^n = a^{mn}$$

The rules for *fractional indices* and *zero index* may also be expressed in general terms as:

$$a^{-n} = 1/a^n \quad \text{and} \quad a^0 = 1$$

The rules of precedence stated in section 1.11 for the manipulation of numbers apply equally to algebraic operations.

Example Multiply $4a^3bc$ by $6b^2c^3$.

Product $= 4a^3bc6b^2c^3$ (add indices of like terms)

$= \underline{24a^3b^3c^4}$

Example Divide $28x^4y^3$ by $-7x^2y$.

Quotient $= \dfrac{28x^4y^3}{-7x^2y}$ (subtract indices of like terms)

$= \underline{-4x^2y^2}$

Example Simplify $(3ab^2)^3.(a^2c)^2$.

Cube each term in the first bracket and square each term in the second:

Expression $= 27a^3b^6.a^4c^2$ (add indices of like terms)

$= \underline{27a^7b^6c^2}$

Example Simplify $4(2x - y) + 6(x + 4y)$.

First expand (multiply out) the brackets:

Expression $= 8x - 4y + 6x + 24y$

Collect like terms:

$= \underline{14x + 20y}$

Example Simplify $3ab\{a + 2(ab + a - 1)\} - b(a + 2)$

Expand the inner bracket first:

Expression $= 3ab(a + 2ab + 2a - 2) - b(a + 2)$

Expand the remaining brackets:

$= 3a^2b + 6a^2b^2 + 6a^2b - 6ab - ab - 2b$

Collect like terms:

$= \underline{9a^2b + 6a^2b^2 - 7ab - 2b}$

Exercise 9 Find the sums of:

1. $x + 2y - 4z$, $2x - y + 3z$ and $2x + 2y + z$
2. $7x - 2xy$, $8x + 7xy$ and $3xy - 4x$
3. $24a - 12b + 2c$, $6a + 2b - c$ and $3b - 4c$

Subtract:

4. $4x + 5y + 2z$ from $7x + 8y + 3z$
5. $2x - 7y + z$ from $3x + 5y - 2z$
6. $5xy + 3yz$ from $5xy - 3yz$

Express the following products in their simplest terms:

7. $3x^3.6x^7$
8. $4x^2y^3.5x^4$
9. $2a^2b.a^5b^2$
10. $(-5x^2)y.7xy^3$

Divide:

11. $15x^7y^3z$ by $5x^6yz$
12. $16a^4x^2y$ by $8axy^3$
13. $64x^2y - 32x^2y^2$ by $8xy$
14. $49a^4b^3 + 28a^3b$ by $7a^3b$

Find the squares of:

15. $3xy^2$
16. $7a^3b$
17. $5x^2y^3z$
18. $-4ab^2c$

Simplify:

19. $\left(\frac{2x^4}{3y^2}\right)^3$ 20. $(-a^2x)^4$ 21. $\left(\frac{1}{3x^2y}\right)^6$ 23. $\left(-\frac{x^3}{2}\right)^5$

Simplify the expressions:

23. $3(2a + 5) - 2(4 - a)$ 24. $2xy\{3(x + y) - 6x^2y\}$

10 PRODUCTS, QUOTIENTS AND FACTORS OF ALGEBRAIC EXPRESSIONS

10.1 Multiplication

Multiplication of algebraic expressions may be carried out in the manner of long multiplication or by extended horizontal presentation.

Consider the product of $x + 4$ and $x + 9$. These are both binomial expressions. They are also expressions of the *first degree*, i.e. they contain only the first power of the unknown.

First method

$$\begin{array}{lrrr} & & x & +\ 9 \\ & & x & +\ 4 \\ \hline \text{Multiply by } x\text{:} & x^2 & +\ 9x & \\ \text{Multiply by 4:} & & 4x & +\ 36 \\ \hline \text{Add:} & x^2 & +\ 13x & +\ 36 \\ \hline \end{array}$$

Second method $(x + 4)(x + 9)$

The multiplication processes are carried out in exactly the same way as before, but are now written in a continuous line.

$$x^2 + 9x + 4x + 36$$

Collect like terms: $\underline{x^2 + 13x + 36}$

With practice the intermediate line can be omitted.

e.g. $(x + 2)(x + 5) = \underline{x^2 + 7x + 10}$

The *first term* is the product of the *first terms* of the binomials.
The *last term* is the product of the *second terms* of the binomials.
The *middle term* is the *algebraic sum of the inner and outer pairs* of the terms in the binomials.

e.g.
$$(x + 3)(x + 6) = x^2 + 9x + 18$$
$$(x + 3)(x - 6) = x^2 - 3x - 18$$
$$(x - 3)(x + 6) = x^2 + 3x - 18$$
$$(x - 3)(x - 6) = x^2 - 9x + 18$$

Note how the combinations of signs affect the products.

In each example given so far the coefficient of x has been unity. Further examples with varying coefficients of x are given below.

$$(4x + 3)(2x + 7) = 8x^2 + 34x + 21$$
$$(2x + 1)(3x - 2) = 6x^2 - x - 2$$
$$(5x - 4)(6x + 3) = 30x^2 - 9x - 12$$
$$(3x - 8)(7x - 1) = 21x^2 - 59x + 8$$

The following examples are products of more complex expressions.

Example

$(4x^2 + 2x - 5)(5x + 2)$

First multiply the first expression by $5x$ and then by 2:

$= 20x^3 + 10x^2 - 25x + 8x^2 + 4x - 10$

$= \underline{20x^3 + 18x^2 - 21x - 10}$

Example

$(x^2 - 7x + 5)(x^2 - 2x + 3)$

$= x^4 - 7x^3 + 5x^2 - 2x^3 + 14x^2 - 10x + 3x^2 - 21x + 15$

$= \underline{x^4 - 9x^3 + 22x^2 - 31x + 15}$

Example

$(3x^3 + 5x^2 - 2x)(3x^2 - 5x - 2)$

$= 9x^5 + 15x^4 - 6x^3 - 15x^4 - 25x^3 + 10x^2 - 6x^3$

$- 10x^2 + 4x$

$= \underline{9x^5 - 37x^3 + 4x}$

Example

$(x^2 + 3xy + y^2)(x - 2y)$

$= x^3 + 3x^2y + xy^2 - 2x^2y - 6xy^2 - 2y^3$

$= \underline{x^3 + x^2y - 5xy^2 - 2y^3}$

DIVISION

Division of compound expressions may be carried out in the manner of long division. The dividend and the divisor should be arranged in descending power of the symbols and arranged as follows:

$$x + 3 \overline{\big)\, 15x^2 + 48x + 9}$$

Divide the first term in the dividend by the first term in the divisor, and write the quotient in the appropriate column *above* the dividend. Multiply the divisor by this quotient, placing the product under the dividend. Subtract, bring down the next term in the dividend, and repeat the process until all terms in the dividend have been used.

$$\begin{array}{r l}
 & \quad 15x + 3 \\
x + 3 \big) & \overline{15x^2 + 48x + 9} \\
 & \underline{15x^2 + 45x} \\
 & \qquad\quad 3x + 9 \\
 & \qquad\quad \underline{\underline{3x + 9}}
\end{array}$$

$\underline{\text{Quotient} = 15x + 3}$

Example Divide $x^4 + 6x^3 + 13x^2 + 12x + 4$ by $x^2 + 3x + 2$.

$$
\begin{array}{r}
x^2 + 3x + 2 \\
x^2 + 3x + 2\,\overline{)\,x^4 + 6x^3 + 13x^2 + 12x + 4} \\
\underline{x^4 + 3x^3 + 2x^2} \\
3x^3 + 11x^2 + 12x \\
\underline{3x^3 + 9x^2 + 6x} \\
2x^2 + 6x + 4 \\
\underline{2x^2 + 6x + 4}
\end{array}
$$

Quotient = $x^2 + 3x + 2$

This result indicates that the dividend is the square of the divisor.

10.2 Factorising

Factors of an algebraic expression may be found by extracting a common factor from the terms or by a process called grouping. The process of factorising is referred to as *resolution* into factors.

FACTORS BY EXTRACTION

Consider the expression $6x^3 + 3x^2 + 3x$. Both 3 and x are common factors of all three terms. $3x$ is the highest common factor (HCF). Divide each term by $3x$, writing the factor outside a bracket containing the resulting quotient.

i.e. $3x(2x^2 + x + 1)$

$3x$ and $2x^2 + x + 1$ are the factors of $6x^3 + 3x^2 + 3x$.

Another example:

$$9x^4 - 3x^3y + 6x^2y^2$$
$$= 3x^2(3x^2 - xy + 2y^2)$$

$3x^2$ is the HCF and $3x^2 - xy + 2y^2$ is the quotient obtained on dividing by $3x^2$.

FACTORS BY GROUPING

If an expression can be arranged in groups that have a compound common factor, the resolution may be made as follows.

$x^2 + xy + xz + yz$

The first two terms have a common factor, x, and the last two terms have a common factor, z. Group the pairs referred to in brackets and extract the common factor from each bracket.

$$(x^2 + xy) + (xz + yz)$$
$$= x(x + y) + z(x + y)$$

The two groups now have a common factor of $(x + y)$, and extracting it gives:

$(x + y)(x + z)$

as factors of the original expression.

Another example:

$$5x + xy + 5y + y^2$$
$$= (5x + 5y) + (xy + y^2)$$
$$= 5(x + y) + y(x + y)$$
$$= \underline{(x + y)(5 + y)}$$

Trinomial expressions of the general form $ax^2 + bx + c$ may have a pair of binomial expressions as factors. We have already seen that the product of two binomials of the first degree is of this trinomial form.

e.g. $(x + 5)(x + 4) = x^2 + 9x + 20$

The reversal of the above multiplication is, of course, the factorisation of the product. Note that the two second terms in the binomial factors (5 and 4) are factors of 20 and that their sum (5 + 4) is the coefficient of x in the middle term of the product.

Consider the expression $x^2 + 7x + 12$. Its factors will be a pair of binomial expressions, the first terms of which will be x. The product of their second terms will be 12 and the algebraic sum of these second terms will be 7, the coefficient of x in the product. Since $3 \times 4 = 12$ and $3 + 4 = 7$, the factors are $(x + 3)$ and $(x + 4)$.

i.e. $\underline{x^2 + 7x + 12 = (x + 3)(x + 4)}$

Example Factorise $x^2 - 5x - 24$.

The second terms of the factors must have a product of −24 and an algebraic sum of −5.

$$-8 \times 3 = -24$$
$$\text{and } -8 + 3 = -5$$

$\therefore$ the factors are $(x - 8)$ and $(x + 3)$

i.e. $\underline{x^2 - 5x - 24 = (x - 8)(x + 3)}$

Example Factorise $x^2 - 12x + 35$.

The second terms of the factors must have a product of 35 and an algebraic sum of −12.

$$-7 \times -5 = 35$$
$$\text{and } -7 + (-5) = -12$$

$\therefore$ the factors are $(x - 7)$ and $(x - 5)$

i.e. $\underline{x^2 - 12x + 35 = (x - 7)(x - 5)}$

Exercise 10 Expand the following products:

1. $(x + 2)(x + 5)$ 2. $(x - 7)(x + 1)$ 3. $(2x + 4)(x + 3)$
4. $(4x - 3)(3x - 4)$ 5. $(x^2 - 6x + 4)(x - 4)$
6. $(a^3 + 3a^2 - 2a + 1)(4a - 5)$ 7. $(a^4 + 2a^2b^2 + b^4)(a^2 + b^2)$

Express the following quotients in their simplest form:

8. $\dfrac{x^3 - x^2 - 9x - 12}{2x^2 + 6x + 6}$

9. $\dfrac{x^9 - y^9}{x^3 - y^3}$

10. $\dfrac{4x^4 - 6x^3 + 3x^2 + 27x + 5}{4x^2 + 6x + 1}$

11. $\dfrac{15a^3 + 10a^2b + 6ab^2 + 4b^3}{3a + 2b}$

Resolve into factors:

12. $x^3 + 2x^2 - x$

13. $ax^3 + bx^2 - cx$

14. $x^2 + xy$

15. $54a^2 - 9a^3 + 27a^4$

16. $19p^4q^3 - 38p^2q^2 + 57p^3q^4$

Find the factors of:

17. $x^2 + x + bx + b$

18. $mn + 4n - pm - 4p$

19. $xy - py + xq - pq$

20. $12x - 4 - 6xy + 2y$

21. $a^2bc + abc^2 - ab^2 - b^2c$

22. $2px + 8py - x - 4y$

Factorise:

23. $x^2 + 10x + 21$

24. $x^2 + 9x + 8$

25. $x^2 + 4xy + 4y^2$

11 EQUATIONS

11.1 General notes on equations

An *equation* is a mathematical statement of the equality of two algebraic expressions. If the statement is true for *any* value of the symbol(s) contained, the equation is an *identical equation*, usually called an *identity*. If it is true for a *particular* value (or limited number of values) of the symbol(s), it is an *equation of condition* and the term 'equation' is generally accepted to refer to this type.

Equations are constructed from known data and are used to arrive at solutions of problems. *Solving* an equation involves finding the value(s) of the unknown(s) represented by the symbol(s) it contains.

$5(x + 3) - 6x + 4$ is an algebraic expression

$5(x + 3) - 6x + 4 = 11x - 5$ is an equation

11.2 Manipulating equations

The manipulation of equations, leading to their solutions, requires that certain basic truths (axioms) are observed. These may be summarised in the statement: *the two sides of an equation will remain equal only if operations on one side are matched by identical operations on the other.*

Examples of this are:

Equals added to equals produce equals
Equals subtracted from equals produce equals
Equals multiplied by equals produce equals
Equals divided by equals produce equals
Equal powers of equals are equal

There are many other examples to complete the list. In everyday terms this can be boiled down to: *what is done to one side of an equation must also be done to the other side.*

Example Solve the equation $7x = 14$.

This is a very simple example. Divide both sides by 7:

$$\frac{7x}{7} = \frac{14}{7}$$

$$\underline{x = 2}$$

The following examples illustrate the application of these basic rules.

11.3 Linear equations

An equation containing only the *first power* of the unknown (or unknowns) is a *simple* or *linear* equation. It may also be referred to as an *equation of the first degree.*

Example

$$5x - 7 = 3x + 4$$

Subtract $3x$ from both sides:

$$5x - 7 - 3x = 3x + 4 - 3x = 4$$

Add 7 to both sides:

$$5x - 7 - 3x + 7 = 4 + 7$$

i.e. $5x - 3x = 4 + 7$

Collect like terms: $2x = 11$

Divide both sides by 2: $\underline{x = 5.5}$

The statement '$x = 5.5$' is the *solution* of the equation. The value 5.5 is the *root* of the equation.

Example

$$3(x + 5) = 4x + 11$$

Expand bracket: $3x + 15 = 4x + 11$

Subtract $3x$ from both sides: $15 = 4x + 11 - 3x$

Subtract 11 from both sides: $15 - 11 = 4x - 3x$

$$4 = x$$

or $\underline{x = 4}$

You will see that in the previous examples the outcome of adding or subtracting equal quantities on both sides has been that the terms move across the 'equals' sign and their signs have changed. Therefore this may be done directly and is called *transposition.*

Example Solve the equation:

$$9x - 5(x + 11) = 3x - 8$$

$$9x - 5x - 55 = 3x - 8$$

Transpose the $3x$ from right to left, changing its sign from + to –, and the 55 from left to right, changing its sign from – to +. The effect is the same as subtracting $3x$ from both sides and adding 55 to both sides. The equation becomes:

$$9x - 5x - 3x = -8 + 55$$

Collect: $\underline{x = 47}$

Example Solve: $3x - 5\{2x - 3(7x - 3) + 6x\} = 5x + 4$

Expand inner brackets:

$$3x - 5(2x - 21x + 9 + 6x) = 5x + 4$$

Expand outer brackets:

$$3x - 10x + 105x - 45 - 30x = 5x + 4$$

Transpose:

$$3x - 10x + 105x - 30x - 5x = 4 + 45$$

Collect: $63x = 49$

Divide both sides by 63: $x = \dfrac{49}{63}$

Cancel by 7: $\underline{x = \dfrac{7}{9}}$

Example

$$\frac{x + 1}{2} = \frac{x - 3}{3}$$

As a first step the fractions should be cleared. This may be done by multiplying both sides by the lowest common multiple of the denominators. This is 6.

$$\frac{6(x + 1)}{2} = \frac{6(x - 3)}{3}$$

$$\therefore 3(x + 1) = 2(x - 3)$$

If this is compared with the original equation it will be seen that the *divisor* of one side has become the *multiplier* of the other side. This process may be applied directly and is called *cross-multiplication.* The numbers move *diagonally* across the equals sign without changing their signs.

Completing the above example, expand brackets:

$$3x + 3 = 2x - 6$$

Transpose:

$$3x - 2x = -6 - 3$$

Collect:

$$\underline{x = -9}$$

Example

$$\frac{2x + 4}{7} = \frac{5x - 9}{4}$$

Cross-multiply the denominators:

$$4(2x + 4) = 7(5x - 9)$$

$$8x + 16 = 35x - 63$$

$$16 + 63 = 35x - 8x$$

$$79 = 27x$$

$$x = \frac{79}{27}$$

$$\underline{x = 2.926}$$

Example

$$\frac{x + 8}{3} - \frac{x + 2}{6} = \frac{x + 3}{4}$$

Multiply both sides by the LCM of the denominators, which is 12.

$$\frac{12(x + 8)}{3} - \frac{12(x + 2)}{6} = \frac{12(x + 3)}{4}$$

$$4(x + 8) - 2(x + 2) = 3(x + 3)$$

$$4x + 32 - 2x - 4 = 3x + 9$$

$$4x - 2x - 3x = 9 - 32 + 4$$

$$-x = -19$$

Multiply both sides by -1:

$$\underline{x = 19}$$

Exercise 11a Solve the equations:

1. $3x - 6 = 18$
2. $2x + 13 = 21$
3. $7x = 5 + 2x$
4. $5(x + 1) = 3(x + 3)$
5. $7(x + 5) = 4x - 10$
6. $12(x - 1) - 3(2x - 3) = 13x + 4$
7. $\dfrac{4x}{5} + \dfrac{x}{4} = 7$
8. $\dfrac{x + 1}{6} = \dfrac{2x - 3}{2}$
9. $\dfrac{4x - 5}{3} - 2(x + 4) = \dfrac{2 - 3x}{5}$

11.4 Construction of equations

The construction of an equation from known data has already been referred to as an approach to solving problems, and the following examples illustrate some typical appiications in other units of the course.

Example The total floor area in the two rooms of a chalet is 28 m² and the larger room has 4 m² more area than the other. Find the area of each.

Let x m² be the area of the larger room. Then $(x - 4)$ m² is the area of the smaller. Total area is $x + (x - 4)$, which is equal to 28 m². The equation is therefore:

$$x + x - 4 = 28$$

$$2x = 28 + 4$$

$$x = \frac{32}{2} = \underline{16\,\text{m}^2}$$

$$x - 4 = \underline{12\,\text{m}^2}$$

The areas are 16 m² and 12 m².

Example The volume of a concrete cube in cubic centimetres is one and a half times the density of the concrete in kilograms per cubic metre. If the cube has a mass of 7.594 kg, what is the size of the cube? Also find the density of the concrete.

Let the edge length of the cube be x cm ($x/100$ m). Then the volume is x^3 cm³ or $(x/100)^3$ m³.

$$\text{Density of concrete} = \frac{x^3}{1.5}\ \text{kg/m}^3$$

$$\text{Density} = \frac{\text{mass}}{\text{volume}}$$

$$\frac{x^3}{1.5} = \frac{7.594}{(x/100)^3}$$

Invert divisor and multiply (on right hand side):

$$\frac{x^3}{1.5} = 7.594 \times \frac{100^3}{x^3}$$

(Note that the index 3 is applied to both numerator and denominator.) Multiply both sides by $1.5x^3$ (the common denominator):

$$x^6 = 7.594 \times 100^3 \times 1.5$$

Take the sixth root of each side:

$$x = \sqrt[6]{(7.594 \times 10^6 \times 1.5)}$$

(This is evaluated by logarithms.)

$$\underline{x = 15\text{ cm}}$$

$$\text{Density} = \frac{15^3}{1.5}$$

$$= \underline{2\,250\text{ kg/m}^3}$$

No.	Log
7.594	0.8804
10^6	6.0000
1.5	0.1761
÷ 6	6)7.0565
15	1.1761
15^3	3.5283
1.5	0.1761
2 250	3.3522

The size of the cube is 15 cm and the density of the concrete is 2 250 kg/m^3.

Example In an electrical circuit three resistances are connected in series to a 9 volt battery, and the current taken from the battery is 1.8 amps. There is a difference of 0.5 ohms between the first and second resistances and a difference of 1 ohm between the second and third. What are the resistances?

Let the first resistance be $x\ \Omega$. Then the second resistance is $x + 0.5\ \Omega$ and the third resistance is $x + 1.5\ \Omega$.

Total resistance $= x + (x + 0.5) + (x + 1.5)\ \Omega$

By Ohm's Law:

$$V = IR \quad (\text{volts} = \text{amps} \times \text{ohms})$$

$$9 = 1.8\{x + (x + 0.5) + (x + 1.5)\}$$

$$9 = 1.8(3x + 2)$$

$$9 = 5.4x + 3.6$$

$$5.4x = 5.4$$

$$\underline{x = 1\ \Omega}$$

Resistances are 1, 1.5 and 2.5 Ω.

Example Two metals are bonded together to form a bimetallic strip. One metal has a coefficient of thermal expansion of 1.2×10^{-5} per °C and the other of 1.6×10^{-5} per °C. When the strip is heated through a temperature range of 40 °C one metal expands by 0.05 mm more than the other. What is the length of the strip?

Let the length of the strip be x mm.
Expansion = $x\alpha t$ (length × coefficient × temperature range)
Expansion of one = $x(1.2 \times 10^{-5})40$ mm
Expansion of the other = $x(1.6 \times 10^{-5})40$ mm

$$x(1.6 \times 10^{-5})40 = x(1.2 \times 10^{-5})40 + 0.05$$

Transpose:

$$x(1.6 \times 10^{-5})40 - x(1.2 \times 10^{-5})40 = 0.05$$

Factorise: $40x \times 10^{-5}(1.6 - 1.2) = 0.05$

Cross-multiply:

$$x = \frac{0.05}{40 \times 10^{-5} \times 0.4} \text{ mm}$$

$$= \frac{0.05 \times 10^{5}}{40 \times 0.4} \text{ mm}$$

$$= \frac{5\,000}{16} \text{ mm}$$

$$= \underline{312.5 \text{ mm}}$$

Example A uniform rod pivoted at its mid-length is in a state of equilibrium when a mass of 320 g is suspended from one end and a mass of 400 g is suspended at a point 40 mm from the other end. How long is the rod?

Let the length of the rod be x mm. The arrangement is sketched in Figure 12. For equilibrium:

Clockwise moments = anti-clockwise moments

i.e. $400\left(\frac{x}{2} - 40\right) = 320\left(\frac{x}{2}\right)$

$$200x - 16\,000 = 160x$$

$$200x - 160x = 16\,000$$

$$40x = 16\,000$$

$$x = \frac{16\,000}{40}$$

$$= \underline{400 \text{ mm}}$$

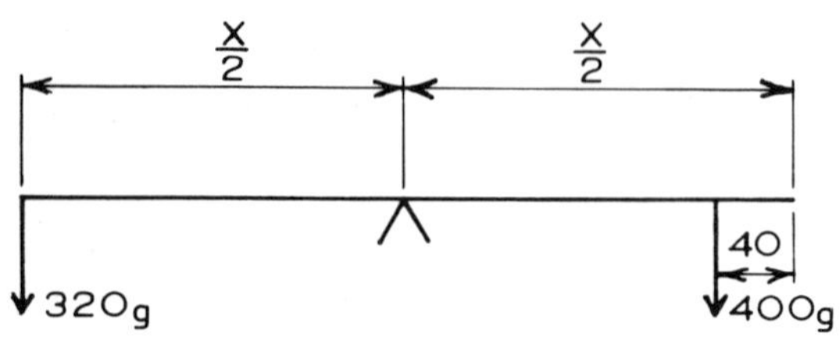

Figure 12

Exercise 11b

1. A rectangle has a length equal to one and two thirds times its breadth and a perimeter of 16 m. What are its dimensions?

2. Ten men employed on a site earn a total of £73.60 daily. The craftsmen earn £8 each per day and the labourers earn £6.40 each per day. How many craftsmen are employed?

3. A uniform rod is pivoted at its mid-point and carries a mass of 1 kg at one end. A 0.5 kg mass is hung at the other end. Where should a 1.5 kg mass be hung in order to produce a state of equilibrium?

4. A uniform timber beam spanning 3 m is simply supported at its ends. The beam weighs 80 N and carries a load of 1 kN at 1 m from the left end and a second load of 1.5 kN that may be moved to any position on the span. Where should the second load be placed in order to make the two support reactions equal?

11.5 Simultaneous equations

An equation connecting two unknowns cannot be solved alone. It is necessary to connect the same two unknowns by a second equation. The two equations may then be solved together, or *simultaneously,* being called *simultaneous equations.* Solution may be by the method of substitution or the method of elimination or a combination of the two.

Consider the following pair of equations:

$$3x + 2y = 9 \qquad \ldots(1)$$

$$x + y = 4 \qquad \ldots(2)$$

Method of substitution

From (2): $y = 4 - x$

Substitute $(4 - x)$ for y in (1): $3x + 2(4 - x) = 9$

$$3x + 8 - 2x = 9$$

$$\underline{x = 1}$$

Substitute 1 for x in (2): $1 + y = 4$

$$\underline{y = 3}$$

Method of elimination

Multiply (2) by 2: $2x + 2y = 8 \qquad \ldots(3)$

$$3x + 2y = 9 \qquad \ldots(1)$$

Subtract (3) from (1): $\underline{x = 1}$

(Note that on subtraction y is eliminated.) Substitute 1 for x in (2), as before:

$$\underline{y = 3}$$

In dealing with simple simultaneous equations with two unknowns it is usually most convenient to find one unknown by the method of elimination and then the other by substitution.

Example

$$3x - 2y = 7 \quad \ldots(1)$$

$$5x + 4y = 30 \quad \ldots(2)$$

Multiply (1) by 2: $6x - 4y = 14 \quad \ldots(3)$

Add (2) and (3): $11x = 44$

$$\underline{x = 4}$$

y is eliminated by the addition. Substitute 4 for x in (1):

$$12 - 2y = 7$$

$$2y = 12 - 7$$

$$\underline{y = 2.5}$$

Example

$$\frac{x}{3} + \frac{y}{2} = 6 \quad \ldots(1)$$

$$\frac{x}{2} + \frac{y}{4} = 5 \quad \ldots(2)$$

Multiply (1) by 6: $2x + 3y = 36 \quad \ldots(3)$

Multiply (2) by 4: $2x + y = 20 \quad \ldots(4)$

Subtract: $2y = 16$

$$\underline{y = 8}$$

Substitute in (4): $2x + 8 = 20$

$$2x = 12$$

$$\underline{x = 6}$$

Example

$$\frac{2}{x} + \frac{3}{y} = 12 \quad \ldots(1)$$

$$\frac{1}{x} + \frac{2}{y} = 7 \quad \ldots(2)$$

It is convenient in this case to find the reciprocals of x and y, which can then be inverted for the solutions. The equations may be written:

$$2\left(\frac{1}{x}\right) + 3\left(\frac{1}{y}\right) = 12 \quad \ldots(1)$$

$$\left(\frac{1}{x}\right) + 2\left(\frac{1}{y}\right) = 7 \qquad \ldots(2)$$

Multiply (2) by 2: $2\left(\frac{1}{x}\right) + 4\left(\frac{1}{y}\right) = 14 \qquad \ldots(3)$

Subtract (1) from (3): $\frac{1}{y} = 2$

$$\therefore \underline{y = \tfrac{1}{2}}$$

Substitute 2 for $\frac{1}{y}$ in (2): $\frac{1}{x} + 4 = 7$

$$\frac{1}{x} = 3$$

$$\therefore \underline{x = \tfrac{1}{3}}$$

It may be necessary to simplify the equations before proceeding with their simultaneous solution.

Example

$$3(x + 2y) - (2x + 7y) = 2 \qquad \ldots(1)$$

$$6x - 8y - 2(x - 3y) = 14 \qquad \ldots(2)$$

Simplify (1): $3x + 6y - 2x - 7y = 2$

$$\therefore x - y = 2 \qquad \ldots(3)$$

Simplify (2): $6x - 8y - 2x + 6y = 14$

$$4x - 2y = 14 \qquad \ldots(4)$$

Multiply (3) by 2: $\underline{2x - 2y = 4}$

Subtract: $2x = 10$

$$\underline{x = 5}$$

Substitute 5 for x in (3): $5 - y = 2$

$$\underline{y = 3}$$

Example

$$x + \frac{y - 4}{5} = \frac{x - 1}{2} \qquad \ldots(1)$$

$$\frac{4y + 1}{3} - \tfrac{1}{2}(3x - 2) = 2y \qquad \ldots(2)$$

Multiply (1) by 10:
$$10x + 2(y - 4) = 5(x - 1)$$
$$10x + 2y - 8 = 5x - 5$$
$$5x + 2y = 3 \quad \ldots(3)$$

Multiply (2) by 6:
$$2(4y + 1) - 3(3x - 2) = 12y$$
$$8y + 2 - 9x + 6 = 12y$$
$$9x + 4y = 8 \quad \ldots(4)$$

Multiply (3) by 2:
$$10x + 4y = 6$$
$$x = -2$$

Substitute in (3):
$$-10 + 2y = 3$$
$$2y = 13$$
$$y = 6.5$$

Exercise 11c Solve the simultaneous equations:

1. $3x + 7y = 23, \quad 5x + 4y = 23$
2. $4x - 3y = 4.5, \quad 2x + 7y = 6.5$
3. $2.5x + 3y = 10, \quad 5x - y = 2.5$
4. $3(x + 3y) - 5(2x - 3) = 7(y + 2),$
   $2x - 2(3x - 2) = 4(y - 12)$
5. $\dfrac{x - 1}{3} + \dfrac{y - 5}{2} = 0, \quad 3y - 4 + \dfrac{x + 2}{3} = 7$
6. $\dfrac{2}{x} + \dfrac{3}{y} = 34, \quad \dfrac{7}{x} + \dfrac{5}{y} = 64$

12 FORMULAE

12.1 Evaluation

An equation that expresses a given quantity (the *subject*) in terms of other, connected, quantities is known as a *formula*. Areas of regular figures and volumes of regular solids are well known examples of the use of formulae.

e.g. area of a circle $A = \pi r^2$

or $A = \frac{1}{4}\pi d^2$

area of a triangle $A = \frac{1}{2}bh$

volume of a sphere $V = \frac{4}{3}\pi r^3$

volume of a cone $V = \frac{1}{3}\pi r^2 h$

(The above formulae are explained in more detail in Chapter F, Geometry).

Having established a formula for any particular set of values or circumstances, the subject may be evaluated after substituting the appropriate values for the other symbols.

Some examples of evaluation from given formulae follow.

Example The volume of a hemisphere is given by the formula $V = \frac{2}{3}\pi r^3$. Find the volume of a hemisphere having a radius of 85 mm. (Use logarithms.)

$$V = \tfrac{2}{3}\pi r^3$$

Substitute: $V = \frac{2}{3}\pi\, 85^3 \text{ mm}^3$

$= \underline{1.286 \times 10^6 \text{ mm}^3}$

No.	Log.
85	1.9294
85^3	5.7882
	0.4972
2	0.3010
	6.5864
3	0.4771
1.286×10^6	6.1093

Note. This may be expressed as 1.286 dm^3 since $1 \text{ dm}^3 = 10^6 \text{ mm}^3$ (dm = decimetre).

Example The time of swing of a simple pendulum is given by $t = 2\pi\sqrt{(l/g)}$. Find the time of swing of a pendulum of length 850 mm. Note that g (the acceleration due to gravity) = 9.81 m/s^2. Use a slide rule.

Rough check:

$$2\pi \triangleq 6$$
$$\sqrt{0.85} \triangleq 0.9$$
$$\sqrt{9.8} \triangleq 3$$
$$6 \times \frac{0.9}{3} = \underline{1.8}$$

$$t = 2\pi\sqrt{\frac{l}{g}}$$

$$= 2\pi\sqrt{\frac{0.850}{9.81}}\text{ s} \quad (l \text{ in metres})$$

Method

Set hairline to 0.85 on A scale (right end).
Move slide until 9.81 on B scale coincides (left end).
Move cursor until hairline is on B scale index.
Transfer to C and D scales (for square root).
Move slide until left index coincides with hairline.
Move cursor until hairline coincides with 2 on C scale.
Move slide until right index on C scale coincides with hairline.
Move cursor until hairline coincides with π on C scale.
Read off result on D scale.
Result = 1.85 s

Example The relative density d of a material may be expressed as:

$$d = \frac{w_1}{w_1 - w_2}$$

where w_1 is the weight of a body of the material in air and w_2 is the apparent weight of the body in water. Find the relative density of lead if a piece weighing 153 g in air appears to weigh only 139.6 g when suspended in water.
Note. Grams are not strictly units of weight, but the term 'weight' is acceptable when masses are used to 'weigh' bodies on a beam balance.

$$d = \frac{w_1}{w_1 - w_2}$$

Substitute: $d = \dfrac{153}{153 - 139.6}$

$$= \frac{153}{13.4} = \underline{11.42} \quad \text{(by slide rule)}$$

Example The power in watts developed when an electric current, I amperes, flows through a resistance, R ohms, is given by $P = I^2R$. Find the power developed when a current of 12.75 A flows through a resistance of 15.3 Ω. Use a calculator.

$$P = I^2R$$

Substitute: $P = (12.75)^2 \times 15.3$ W

Sequence of operations:

[1] [2] [·] [7] [5] [×] [×] [1] [5] [·] [3] [=]

Result = $\underline{2\,487.2 \text{ W}}$ (or 2.487 kW)

12.2 Changing subject of formula

Standard formulae may require rearranging before use. Changing the subject of a formula requires the application of the same basic processes that are applied when manipulating equations. The following examples illustrate this.

Example Rearrange the formula $A = \pi d^2/4$ to make d the subject, and then find the diameter of the circle that has an area of 3 530 mm^2.

$$A = \frac{\pi d^2}{4}$$

Multiply both sides by $4/\pi$ (i.e. cross-multiply the 4 and the π):

$$\frac{4A}{\pi} = d^2$$

Reverse: $d^2 = \dfrac{4A}{\pi}$

Take $\sqrt{\ }$ of both sides:

$$d = \sqrt{\frac{4A}{\pi}}$$

$$= 2\sqrt{\frac{A}{\pi}} \quad (\text{since } \sqrt{4} = 2)$$

Substitute: $d = 2\sqrt{\dfrac{3\,530}{\pi}}$

$$= \underline{67.04 \text{ mm}} \quad (\text{correct to 4 significant figures})$$

Example The increased length, l, of a wire initially of length l_o, when heated through t°C is given by $l = l_o(1 + \alpha t)$, where α is the coefficient of linear thermal expansion. Change the subject to t.

$$l = l_o(1 + \alpha t)$$

Divide both sides by l_o (or cross-multiply l_o):

$$\frac{l}{l_o} = 1 + \alpha t$$

Subtract 1 from both sides (or transpose the 1):

$$\frac{l}{l_o} - 1 = \alpha t$$

Divide both sides by α (or cross-multiply) and reverse:

$$\underline{t = \frac{1}{\alpha}\left(\frac{l}{l_o} - 1\right)}$$

Example The load-carrying capacity in newtons of a tie-bar with a rectangular cross-section $(b \times t)\,\text{mm}^2$ and perforated by two rivet holes each of diameter d mm is given by the formula $W = f_t(b - 2d)t$, where f_t is the permissible stress in the bar in N/mm^2. Change the subject to b and find the necessary breadth of such a tie-bar 12 mm thick to carry 350 kN if the rivet holes are 22 mm diameter and $f_t = 150\ \text{N/mm}^2$.

$$W = f_t(b - 2d)t$$

Divide both sides by $t.f_t$:

$$\frac{W}{t.f_t} = b - 2d$$

Add $2d$ to both sides and reverse:

$$b = \frac{W}{t.f_t} + 2d$$

$$W = 350\,\text{kN}$$
$$= 350 \times 10^3\,\text{N}$$

Substitute (note that W is in newtons):

$$b = \frac{{}^{7}\cancel{350} \times 10^3}{{}_{3}\cancel{150} \times 12} + 2(22)\,\text{mm}$$

$$= \frac{7\,000}{36} + 44\,\text{mm}$$

$$= 194.4 + 44\,\text{mm}$$

$$= \underline{238.4\,\text{mm}} \quad \text{(say 240 mm)}$$

Example The velocity, v m/s, of a body t seconds after moving from an initial velocity of u m/s with a uniform acceleration of a m/s^2 is given by the formula $v = u + at$. Change the subject to a, and find the acceleration of a body that increases its velocity from 3.5 m/s to 4.75 m/s in 2.2 seconds.

$$v = u + at$$

Subtract u from both sides (transpose u):

$$v - u = at$$

Divide both sides by t and reverse:

$$a = \frac{v - u}{t}$$

Substitute:

$$a = \frac{4.75 - 3.5}{2.2}\,\text{m/s}^2$$

$$= \frac{1.25}{2.2}\,\text{m/s}^2$$

$$= \underline{0.568\,\text{m/s}^2}$$

Example The quantity of heat, Q joules, required to raise the temperature of a copper calorimeter and the water it contains through a range of t °C is given by:

$$Q = (m_W \cdot c_W + m_C \cdot c_C)t$$

where m_W = mass of water in calorimeter in kg
c_W = specific heat capacity of water in joules per kg °C
m_C = mass of calorimeter in kg
c_C = specific heat capacity of copper in joules per kg °C

Rearrange the formula to make c_C the subject, and find its value if it takes 158.4 kJ to heat a copper calorimeter having a mass of 250 g and containing 450 g of water from 20 °C to 100 °C. Note that c_W = 4180 J/kg °C.

$$Q = (m_W \,.\, c_W + m_C \,.\, c_C)t$$

Divide both sides by t:

$$\frac{Q}{t} = m_W \,.\, c_W + m_C \,.\, c_C$$

Subtract $m_W \,.\, c_W$ from both sides and reverse:

$$m_C \,.\, c_C = \frac{Q}{t} - m_W \,.\, c_W$$

Divide both sides by m_C:

$$\underline{c_C = \frac{1}{m_C}\left(\frac{Q}{t} - m_W \,.\, c_W\right)}$$

Substitute:

$$c_C = \frac{1}{0.25}\left(\frac{158.4 \times 10^3}{100 - 20} - 0.45 \times 4\,180\right) \text{ J/kg °C}$$

$$= 4(1\,980 - 1\,881) \text{ J/kg °C}$$

$$= \underline{396 \text{ J/kg °C}}$$

Exercise 12

1. The pressure in pascals (N/m^2) at the depth h metres in a fluid of density ρ kg/m^3 is given by:

$$\rho = hdg$$

Find the pressure at a depth of 1.35 m below the surface of a fluid having a density of 1 087 kg/m^3. Use logarithms and give the result in standard form. (g = 9.81 m/s^2.)

2. The formula:

$$V = 2f_S \tfrac{1}{4}\pi d^2$$

gives the strength value of a rivet in double shear. Find the strength of a 22 mm diameter (d) rivet in double shear, given that f_S = 100 N/mm^2. Give the result in kN.

3. The modulus of elasticity of a material in kN/mm^2 is expressed in the formula:

$$E = \frac{Wl}{Ax}$$

where W is the load in kN that will produce an elongation of x mm in a body of length l mm and a cross-sectional area of A mm^2. Find E when W = 20 kN, l = 205 mm, A = 210 mm^2 and x = 0.093 mm.

4. The velocity, v m/s, of a body after moving through a distance s metres with uniform acceleration a m/s^2 from an initial velocity u m/s is given in the formula:

$$v^2 = u^2 + 2as$$

Find the velocity of a body that moves from an initial velocity of 12.6 m/s through a distance of 23 m with an acceleration of 8.75 m/s^2.

5. The velocity ratio of a differential pulley block is given by the formula:

$$\text{V.R.} = \frac{2R}{R - r}$$

where R and r are the radii of the larger and smaller portions of the stepped pulley. Find the velocity ratio of such a pulley block having diameters of 120 mm and 100 mm.

6. $K = mv^2$ gives the kinetic energy in joules of a body of mass m kg moving with a velocity of v m/s. Find the kinetic energy at ground level of a body that has been dropped from a height of 20 m if its mass is 16 kg and its velocity is given by the formula $v^2 = 2gh$, h being the distance in metres through which it has travelled from rest. ($g = 9.81$ m/s^2.)

7. The modulus of elasticity of a material in kN/mm^2 determined from results of a tensile test is given by the formula:

$$E = \frac{Wl}{Ax}$$

where W = force applied to specimen
A = cross-sectional area of specimen
l = gauge length
x = extension on gauge length

Change the subject to x, and find the elongation produced in a steel specimen of cross-sectional area 156 mm^2 and gauge length 150 mm by a force of 26.2 kN. (E for steel = 210 kN/mm^2.)

8. The formula:

$$R_t = R_0(1 + \alpha_0 t)$$

gives the electrical resistance in ohms of a body at t °C in terms of its resistance at 0 °C (R_0 ohms) and its temperature coefficient of resistance at 0 °C (α_0). Change the subject to t, and calculate the temperature range through which a copper wire having a resistance of 2.5 ohms at 0 °C must be heated to increase its resistance by 0.5 ohms. (α_0 for copper = 0.004.)

9. The volume of a cone is expressed in terms of the radius of its base and its vertical height by the formula:

$$V = \tfrac{1}{3}\pi r^2 h$$

Change the subject to r, and find the diameter of the base of the cone of height 400 mm that has a volume of 9.426 dm^3.

10. The surface tension, T N/m, of a liquid may be determined from the height, h m, that the liquid will rise in a capillary tube of internal diameter d m, and is given by:

$T = \frac{1}{4}\rho hdg$

where ρ is the density of the liquid in kg/m^3. Find the height to which water will rise in a tube of 0.556 mm bore if the surface tension of water is 72.679 mN/m. (Density of water = 1 000 kg/m^3; $g = 9.81\ m/s^2$.)
Note. The surface tension of a liquid varies with temperature, and the above value relates to 20 °C.

SUMMARY OF C

Symbolic expression: the use of letters to represent numbers.
Laws of indices:

$$a^m \times a^n = a^{m+n}$$

$$a^m \div a^n = a^{m-n}$$

$$(a^m)^n = a^{mn}$$

$$a^{-n} = 1/a^n$$

$$a^0 = 1$$

Factors: if an algebraic expression is the product of a number of terms (or expressions), then each is a factor of that expression.
Equation: treat both sides of an equation identically in order to retain their equality.
Linear equation: an equation containing only the first power of the unknown(s).
Simultaneous equations: a number of equations (equal to the number of unknowns) that are solved together.
Formula: an equation in which one quantity (the *subject*) is expressed in terms of connected quantities. Rearrangement to change the subject requires that both sides be treated identically.

REVISION EXERCISE C

Add together the following expressions:

1. $8p - 7q + 4r$, $6q - 5p - 6r$ and $5r + 4q + 11p$
2. $7s - 8t$, $3r + 4t$, and $5r + 2s$

Subtract:

3. $-11pq + 5rs$ from $10qr + pq - 2rs$
4. $x^3 + 4x^2 - 2x + 1$ from $x^3 - 2x^2 + 2x - 1$

Simplify:

5. $7p^4q^7r^3 . 4p^2qr^5$
6. $-3a^2 . bc^4 . abc$

Divide:

7. $12a^3b^2c$ by $4a^2b^3c^2$
8. $18x^3y + 12x^2y^3 - 6x^3y^3$ by $3xy^2$

Simplify:

9. $\left(\frac{1}{3xy}\right)^2$
10. $\left(-\frac{3ab^2}{4x^2y}\right)^2$
11. $(7p^2qr^3)^2$
12. $(\frac{2}{3}x^2y^3)^3$
13. $3(3x^2 + 7x - 1) + x(x^2 - 5)$
14. $4(x - y) - 6[x - 4\ \{1 - 7(x - y)\}]$

Find the products of the following expressions:

15. $(x + 8)(x + 8)$
16. $(3x - 1)(2x + 2)$
17. $(x^2 + 5)(x - 4)$
18. $(3x^2 + 7xy - y^2)(x + 3y)$
19. $(\frac{1}{2}x^2 + 3x - \frac{1}{3})(\frac{1}{2}x + \frac{1}{3})$

Find the following quotients:

20. $x^8 - y^8 \div x^3 + x^2y + xy^2 + y^3$
21. $3x^3 - 4x^2y - 3xy^2 - 2y^3 \div 3x^2 + 2xy + y^2$

Factorise the following expressions:

22. $45x^2y - 225xy^2$
23. $3abc + 9a^2b^2c^2$
24. $121x^5y - 11x^3y^2 + 55xy^5$
25. $x^2 - xy + 3x - 3y$
26. $2x^2y + xyz - 2xy^2 - y^2z$
27. $3ax - ay + 6bx - 2by$
28. $p^3q - p^3 + pq^3 - pq^2$

Solve the following equations:

29. $x - 17 = 3 - 9x$
30. $5x + 2 = x + 8$
31. $6(x + 1) + 3(2x - 5) = 9(x - 3) + 3(3x + 2)$
32. $7x - (3x + 7) + 11(x - 5) + 4(x + 1) = 3x - 1$
33. $\frac{x}{6} + \frac{y}{7} = 4.5, \quad \frac{x}{5} + y = 17$
34. $\frac{5}{x} - \frac{2}{y} = 8, \quad \frac{3}{x} + \frac{4}{y} = 36$
35. $\frac{x - 3}{4} + \frac{3x + 1}{6} = \frac{x + 2}{3}$
36. $\frac{x}{2} + \frac{x}{3} - \frac{x}{4} = 14$

37. Three electrical resistances connected in series have a combined value of 4.6 ohms. The first is one third of the second and the third is two ohms less than the difference between the other two. Find the value of each.

38. A rectangular plot has its longer side 28 m greater than the shorter side. If the length were reduced by 34 m and the breadth increased by 40 m the area would remain unchanged. Find the plot's dimensions.

39. A man spent £24 750 on business premises, which sum included repairs and adaption for use as a warehouse. If the cost of repairs was a tenth of the purchase price and the handling equipment installed cost one eighth of the total spent on purchase and repairs, calculate the purchase price.

40. The distance, s metres, travelled by a body moving with uniform acceleration a m/s^2 for t seconds from an initial velocity u m/s is given by:
$s = ut + \frac{1}{2}at^2$
Find s when $u = 3.5$ m/s, $t = 6$ s and $a = 5.25$ m/s^2.

41. The percentage absorption by volume of a brick is given by the formula:

$$A_v = \frac{b - a}{b - c} \times 100\%$$

where a = weight of the brick when dry
b = weight of the brick when saturated
c = weight in water of the saturated brick
Find the percentage absorption by volume of a specimen brick for which the above values are: $a = 3.25$ kg, $b = 3.66$ kg, $c = 2.23$ kg.

42. The support reactions, R_1 and R_2, of a simply supported beam spanning l metres and loaded with a uniformly distributed load of w kN/m over the left half of its span plus a concentrated load of W kN at the mid-point of the other half are given by:

$$R_l = \tfrac{1}{8}(2W + 3wl)$$

and $$R_r = \tfrac{1}{8}(6W + wl)$$

Find the reactions when the span is 3.75 m, the uniformly distributed load is 15 kN/m and the concentrated load is 10 kN.

43. The compressive stress in N/mm^2 developed in a punch used to produce a circular blank from steel plate is given by:

$$f = \frac{4F}{\pi d^2}$$

where F is the force applied in newtons and d is the diameter of the blank in mm. Change the subject to d, and find the maximum diameter blank that can be punched out by a force of 275 kN using a punch in which the limiting compressive stress is 240 N/mm^2.

44. The resistivity, ρ, of a material in $\mu\Omega$ m (microhm metres) is given by the formula:

$$\rho = \frac{RA}{l}$$

where R is the resistance in $\mu\Omega$ of a body of the material having a cross-sectional area of A m^2 and a length l m. Change the subject to l, and find the length of a constantan wire of diameter 0.3 mm required to wind a 12 Ω resistance coil. (ρ for constantan = 0.5$\mu\Omega$ m.)

45. The support reactions, R_L and R_R, of a beam simply supported over a span of l m and carrying a central point load W kN plus a uniformly distributed load over the whole span of w kN/m are given by:

$$R_L = R_R = \tfrac{1}{2}(W + wl)$$

Rearrange the formula to make w the subject, and find its value when $l = 3$ m, $W = 12$ kN and $R_L = 12.5$ kN.

46. The force in newtons required to punch a hole of diameter d mm in a steel plate of thickness t mm is given by:

$$F = f_s \pi d t$$

where f_s is the shearing stress for the steel in N/mm^2. Change the subject to f_s, and find the shearing stress value if a force of 350 kN is required to punch a hole 100 mm diameter in a plate 4.5 mm thick.

47. The velocity, v m/s, of a body t seconds after moving from an initial velocity u m/s with a uniform acceleration a m/s^2 is given by:

$$v = u + at$$

Change the subject to a, and find the acceleration of a body that increases its velocity from 3.5 m/s to 4.75 m/s in 1.2 seconds.

48. The total surface area of a solid cylinder is expressed in the formula:

$$S = 2\pi r(h + r)$$

Find the height (h) of a cylinder of diameter 0.75 m that has a surface area of 2.75 m^2.

D Graphs and mappings

This chapter deals with:
Diagrammatic representation
Conversion graphs, adjacent and non-adjacent scales
Rectangular axes
One to one correspondence, proportionality
Presentation, layout of graphs; scales, labelling
Cartesian coordinates, plotting a point
Linear laws
Gradient of a line

13 DIAGRAMMATIC REPRESENTATION

Pictorial representation of the relationships between sets of values allows quick and easy interpretation. Much information can be presented in a small space, and mental assessments may be made more quickly than would be possible from the same details expressed in words and numbers. A variety of pictorial and diagrammatic methods is possible, some of which are indicated here. Others are shown in Chapter E, Statistics.

A graph is a line, straight or curved, that represents the relationship between two quantities. In order to plot a graph it is necessary to know this relationship, either as a *mathematical law* or as *sets of corresponding values*.

13.1 Adjacent scales

Conversion of units from one system to another is a process that is conveniently carried out graphically, and in its simplest form it is achieved by constructing *adjacent scales* on a single line. Figure 13 shows such a conversion between inches and millimetres. The two scales are marked along opposite sides of the line with their zero marks coinciding. The set of millimetres is *mapped* on to the set of inches.

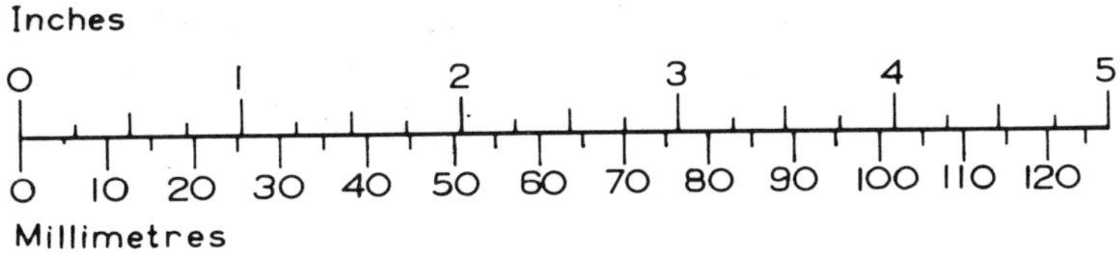

Figure 13

In the case of non-linear quantities, such a relationship may not be so easily visualised. Suppose we wish to convert cubic yards to cubic metres over a range of 10 yd^3.

$1 \text{ yd}^3 = 0.7646 \text{ m}^3$

The adjacent scales are shown in Figure 14. Select a suitable scale to represent up to 10 yd^3 and divide the line AB accordingly.

$10 \text{ yd}^3 = 10 \times 0.7646 \text{ m}^3 = 7.646 \text{ m}^3$

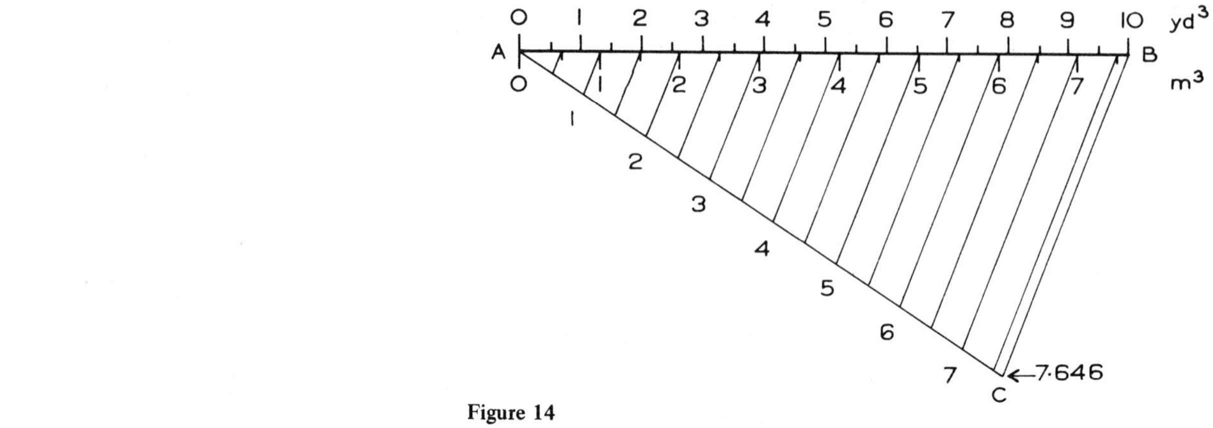

Figure 14

Choose a second scale to represent 7.646 m^3 on a length approximating to AB, and mark this scale along AC at between 30° and 40° to AB. Join BC and draw parallels to it from the divisions on AC to map a proportional scale on to AB. The two scales now form a conversion graph on AB.

There is usually little point in plotting conversion graphs beyond the range of 10 units, since greater values may be taken as multiples of those within that range. In the above example, if we wish to convert 63 yd^3 it may be written as 6.3 × 10 yd^3. From the graph:

$$6.3\ \text{yd}^3 = 4.82\ \text{m}^3$$

$$\therefore\ \underline{63\ \text{yd}^3 = 48.2\ \text{m}^3}$$

13.2 Non-adjacent scales

Conversion scales may also be constructed on a pair of parallel lines (non-adjacent scales). Consider the relationship between pounds and kilograms up to 10 kg.

$$1\ \text{kg} = 2.2\ \text{lb}$$

$$\therefore\ 10\ \text{kg} = 22\ \text{lb}$$

Choose a suitable scale and mark off divisions on a vertical line (AB in Figure 15) to represent the required range of kilograms, reading upwards. Choose a second scale on which 22 lb may be represented by a length approximating to AB. Use this scale to mark off divisions up to 22 lb on a line parallel to AB, this time reading downwards as shown at CD. Join the zero marks of the two scales (0, 0) and join the maximum values (10 kg, 22 lb), marking their point of intersection P. Any straight line through P between the two scales will link corresponding values.

Two conversion examples are shown on the diagram. It is advisable to space the two scales at such a distance apart that the lines joining their extremes are at angles of not less than 30° to the vertical.

13.3 Conventional graphs

A more conventional approach to graphical work involves the use of a pair of reference lines drawn at right angles to each other called *axes*. The two quantities being related are plotted on scales marked along these axes, and corresponding pairs of values are used to plot points on the graph.

Figure 16 shows the use of such a graph for the conversion of miles per gallon to kilometres per litre up to 40 miles per gallon.

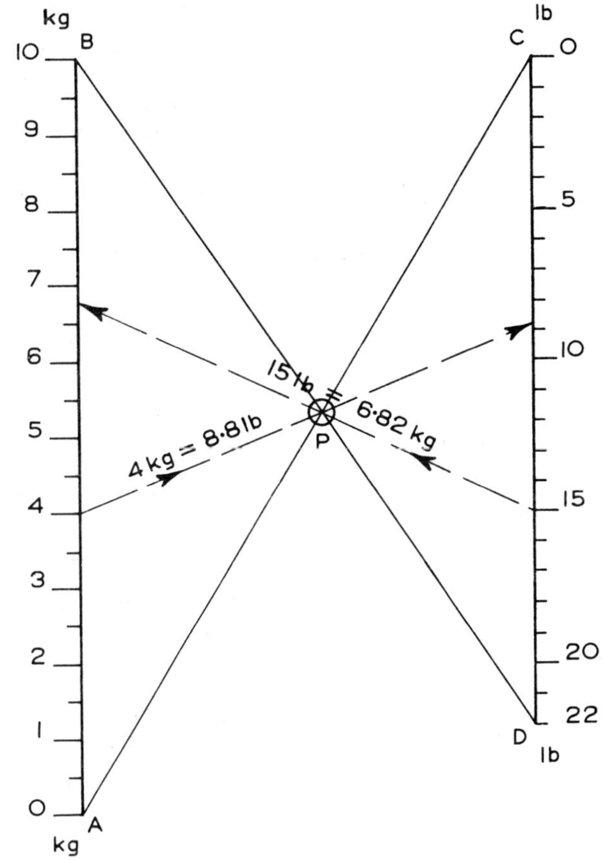

Figure 15

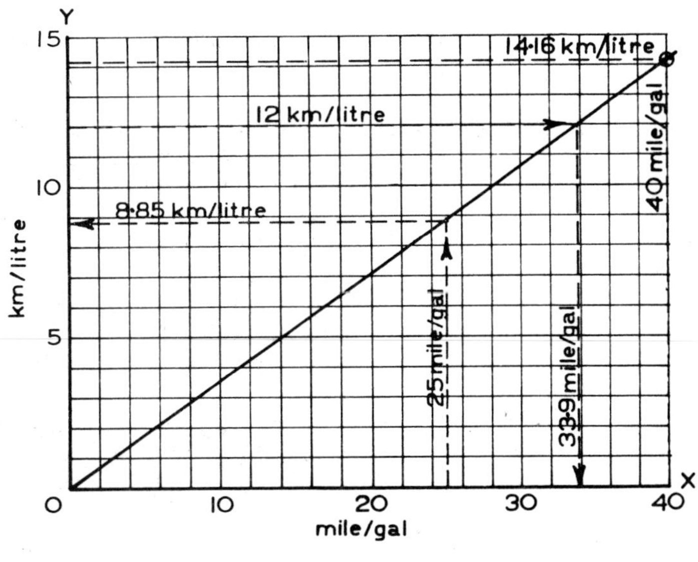

Figure 16

1 mile/gal = 0.354 km/litre

∴ 40 mile/gal = 14.16 km/litre

A scale of mile/gal is marked along the horizontal axis, 0X, and a scale of km/litre along the vertical axis, 0Y. A straight line is drawn from their point of intersection, 0, to the point where the horizontal line from 14.16 km/litre meets the vertical line from 40 mile/gal. This line is the graph from which conversions may be made. Two examples are shown in the figure.

Exercise 13a Construct adjacent scales for the following conversions:

1. Square yards to square metres. 1 yd^2 = 0.836 m^2. Use the scales to convert 280 yd^2 to square metres and 75 m^2 to square yards.
2. Ounces to grams up to 1 lb. 1 oz = 28.35 g. Convert 9 oz to grams and 25 g to ounces.

Construct non-adjacent scales to make the following conversions:

3. Density in pounds per cubic foot to kilograms per cubic metre. 1 lb/ft^3 = 16.02 kg/m^3. Convert 144 lb/ft^3 to kilograms per cubic metre and 7800 kg/m^3 to pounds per cubic foot.
4. Gallons to litres. 1 gal = 4.546 litres. Convert 7.8 gal to litres and 35 litres to gallons.

Draw conversion graphs on a system of rectangular axes, 0X and 0Y, for the following:

5. Tons per square inch to newtons per square millimetre. 1 $tonf/in^2$ = 15.44 N/mm^2. Convert 7.5 $tonf/in^2$ to newtons per square millimetre and 150 N/mm^2 to tonf per square inch. (Tonf means ton force and is used because the ton is strictly a unit of mass.)
6. British thermal units per pound degree Fahrenheit to kilojoules per kilogram degree Celsius. 1 Btu/lb °F = 4.187 kJ/kg °C. Convert 0.095 Btu/lb °F to kJ/kg °C and 2.5 kJ/kg °C to Btu/lb °F.

13.4 Mapping

In each of the examples so far shown, it is apparent that the quantities represented are so related that for each value in one set there is one, and only one, corresponding value in the other set. This is called *one to one correspondence*, and the plotting of such relationships is *one to one mapping*.

13.5 Proportionality

Any proportional relationship between quantities will produce a *straight line* when plotted as a graph. Such relationships include those between the following quantities:

i) load applied and extension produced for an elastic material,
ii) distance travelled and time taken for a body moving with uniform velocity,
iii) current and potential difference for a conductor at constant temperature,
iv) pressure and depth in a static fluid.

Example A copper wire was subjected to tensile loading in stages of 10 N up to a maximum of 60 N. The resulting extensions are shown in the table below. Plot a graph of load against extension, and from it estimate:

i) the extension resulting from a load of 37.5 N,
ii) the load required to produce an extension of 0.2 mm.

Load (N)	0	10	20	30	40	50	60
Extension (mm)	0	0.045	0.090	0.135	0.180	0.225	0.270

The graph is shown in Figure 17. The fact that the points plotted lie on a straight line indicates that extension is proportional to the load producing it (*Hooke's Law*). From the graph we see that the required values are:
i) extension resulting from a load of 37.5 N = 0.169 mm,
ii) load producing an extension of 0.2 mm = 44.4 N.

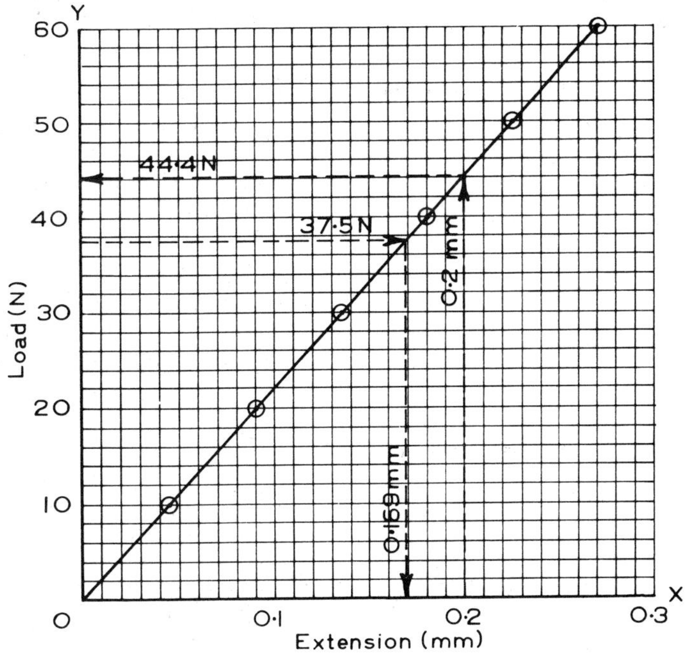

Figure 17

Exercise 13b

1. A certain size of steel beam section has a mass of 45 kg per metre of its length. Construct a graph to read the mass of any length of the section up to 10 m. From the graph find the mass of a 7.75 m length and the length of a beam having a mass of 380 kg.

2. The table shown below gives corresponding values of potential difference and current measured for a resistance of 1.5 ohms. Plot a graph to show the current passing for a range of P.D. values up to 10 volts.

P.D. (volts)	1	3	5	7
Current (amps)	0.67	2.00	3.33	4.67

From the graph determine the current passing when the P.D. is 9.5 volts and the voltage when the current is 4 amps.

3. Using the corresponding values of temperature represented by (32 °F, 0 °C) and (212 °F, 100 °C), draw a graph to convert tempera-

tures ranging from 0 °F to 300 °F. Use the graph to complete the following table.

°F	10	50		150		
°C			30		80	120

4. The shank diameters of wood screws between gauge number 1 and gauge number 10 vary directly as the gauge number. Given that the shank diameter of a No. 1 screw is 1.65 mm and that of a No. 10 screw is 4.80 mm, construct a graph to give, by interpolation, the shank diameters of all intermediate gauge numbers. From the graph read off the shank diameters of No. 4 and No. 8 screws. What size screw would you select for use in a 4.5 mm diameter hole?

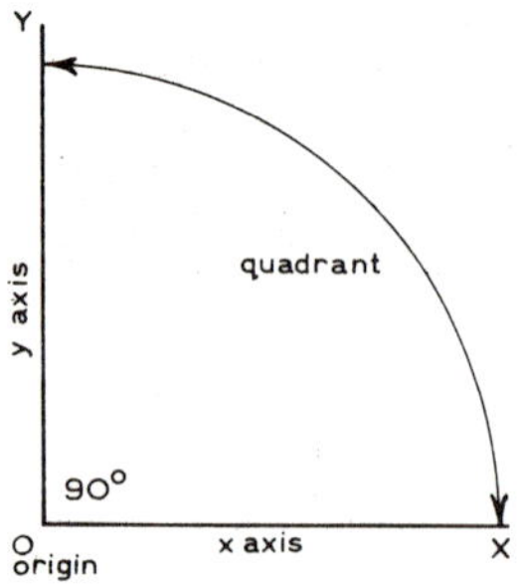

Figure 18

13.6 Rectangular axes

The reference lines 0X and 0Y used in the previous two examples are called rectangular axes because they are at right angles to each other. Their point of intersection, corresponding to the zero marks of both scales, is called the *origin* of the reference system they form. (It is important to note that when the point of intersection does not coincide with *both zero marks* it is not the origin.)

The relationship between any pair of corresponding values on the two axes may be defined by a point plotted, or mapped, in the space (quadrant) between them. 0X is conventionally referred to as the *x axis* and 0Y as the *y axis*. See Figure 18.

13.7 Scales for plotting

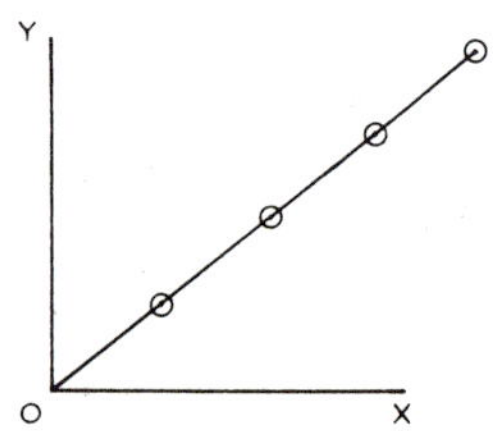

Figure 19

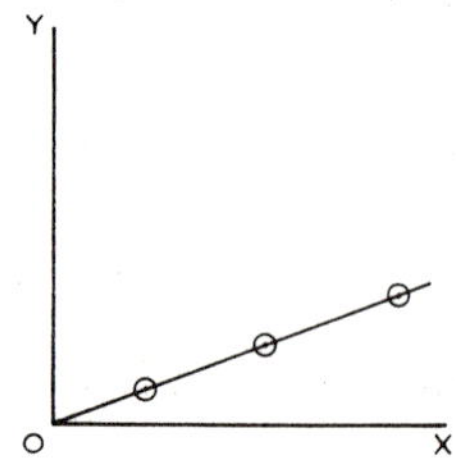

Figure 20

The scales used to represent the quantities plotted should be chosen with care. In general, the larger the scale the more accurate will be the results. It is not advisable, however, to arrange a graph so that it extends right to the edge of the paper, since a border is required in which to label the scale and describe the quantity represented. Much of the advantage of pictorial presentation is lost if the diagram is not arranged neatly on the paper.

For most graphical work squared paper is preferred, and this will normally be divided in multiples of ten. A useful ruling is that of 2 cm squares divided into tenths. Closer ruling of 1 cm squares divided into tenths give such a fine grid that it is difficult to use for normal pencil work unless the ruling is feint.

Scales based on multiples of 2, 5 or 10 are best. Multiples of 3 should be avoided, since the resulting division of squares would lead to inaccuracies in both plotting and reading. It is not necessary to use the same scale on both axes, although it is sometimes convenient to do so. The range to be marked along each axis needs careful thought when planning the layout of the graph. An insufficient range will lead to the situation shown in Figure 19, whereas too great a range will result in an inadequate graph of the type shown in Figure 20. Both should be avoided.

13.8 Presentation

Both axes should be clearly labelled with the quantity represented and the units used. A suitable method is to write the quantity along the axis in block letters followed by the appropriate units in a bracket. They should be placed sufficiently clear of the axis to enable the scale to be marked. Avoid overcrowding of the figures on the scales, and keep them just clear of the line for clarity.

13.9 Coordinates

Any point in the quadrant between the x and y axes may be defined by its perpendicular distances from the axes. In Figure 21 the point P is at distance 2 units from the y axis and distance 3 units from the x axis. Its position is conveniently described as (2, 3). Similarly the point Q is at (5, 6). It is conventional to write the horizontal distance from the y axis first, followed by the vertical distance from the x axis. In other words the point is defined in the terms (x, y). The x value is called the *abscissa* of the point and the y value its *ordinate*. Together they are the point's *Cartesian coordinates*.

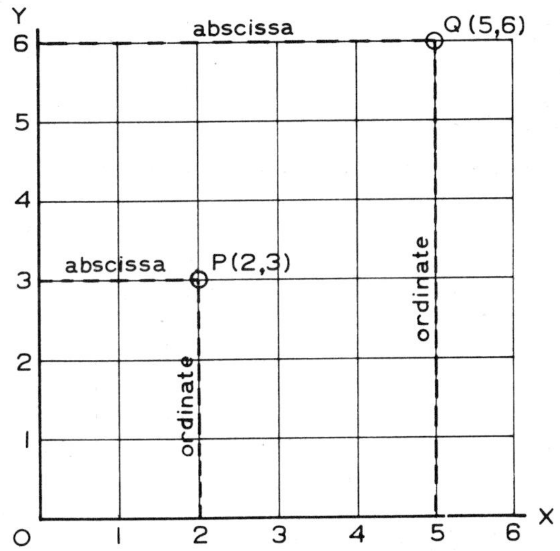

Figure 21

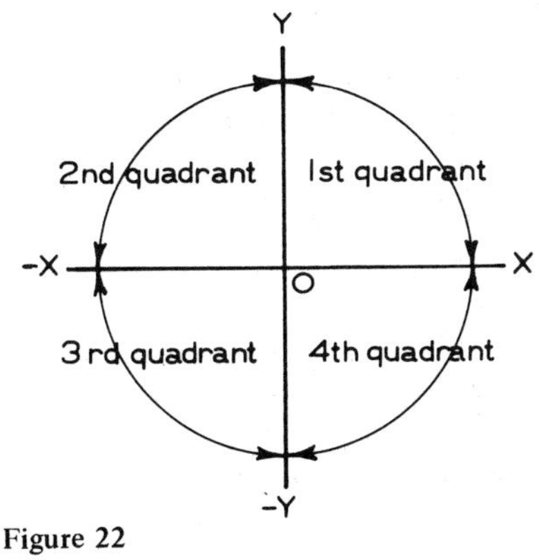

Figure 22

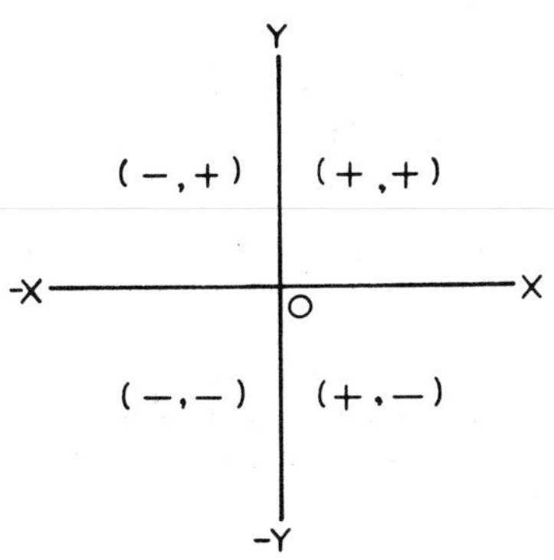

Figure 23

Accurate plotting is essential, and when its position is determined each point should be defined by a clear mark. An encircled point gives clarity (⊙). Alternatively a small neat cross is acceptable (+). A sharp-pointed pencil should always be used; never use a ballpoint pen for graphical work. A convenient method of plotting is to find the value of the abscissa along the x axis, and then mark the corresponding y value up the ordinate at that point. You will then be using the values in their conventional order of description.

Although practical problems involving physical quantities such as length, mass, density and so on require the use of *positive* values only, it is often necessary to deal with *negative* quantities in graphical work. For this purpose the axes are extended to the left of the origin and below it as shown in Figure 22. The two axes now embrace four *quadrants,* which are conventionally numbered as shown.

Values plotted to the right of the y axis are positive and values plotted to the left are negative. Values plotted above the x axis are positive and values plotted below are negative. Figure 23 shows the resulting signs of coordinates of plotted points.

Example Plot the points that have the coordinates (5, 3), (–2, 3), (–2, 4) and (5, –4), and show that a square is formed when they are joined.

Examination of the signs of the coordinates shows that each is in a different quadrant. Figure 24 shows the points plotted and joined. Two pairs of parallel lines result, one pair horizontal since the lines join ordinates of equal value, the other pair vertical since they join abscissae of equal value. The distance between the vertical lines is 7 units (from −2 to 5) and the distance between the horizontal lines is 7 units (from −4 to 3), and therefore the figure formed is a square.

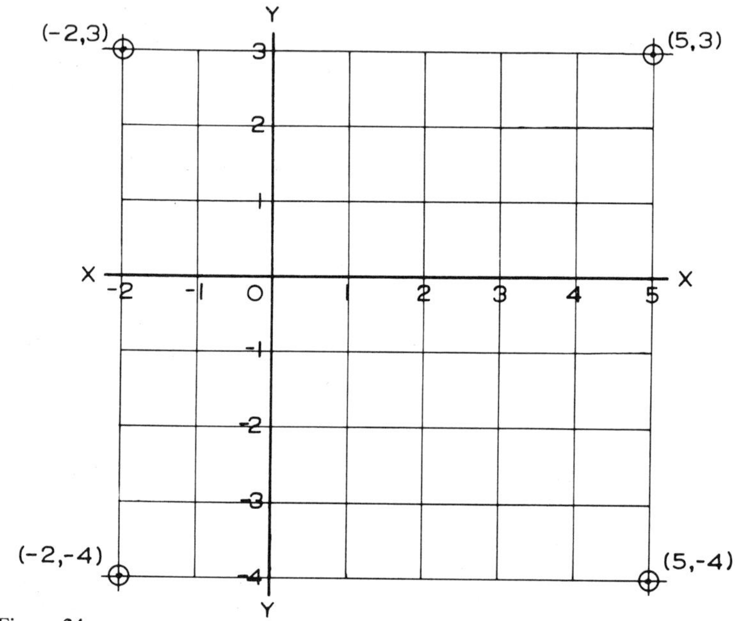

Figure 24

13.10 Plotting straight-line graphs

We saw in section 11.3 that an equation of the *first degree* is also referred to as a *linear equation*. This is because the graph of two variable quantities connected by such an equation will always be a straight line. The equation may be called the *linear law* of the relationship.

Example Plot a graph of the relationship $y = 5x$ between values of x from 0 to 10.

Clearly, whatever value we give to x, y will be five times that value. A table of corresponding pairs of values is easily constructed.

x	0	2	4	6	8	10
y	0	10	20	30	40	50

Figure 25 shows the mapping of these sets. As previously stated, we really require only two points (plus a third to guard against error) to draw the line since x is proportional to y. Note the linear law marked on the line.

The following examples show graphs of other linear equations.

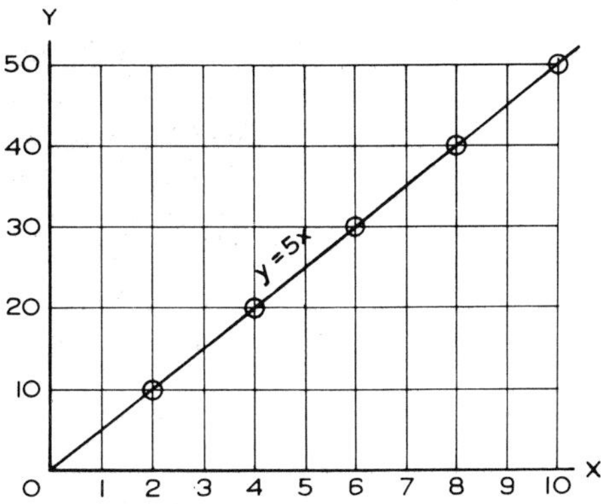

Figure 25

Example Plot the graph of $2y = 5x + 1$ for values of x from -4 to 6.

To simplify the calculation of corresponding values, divide both sides of the equation by 2.

$y = 2.5x + 0.5$

This time, take only three values of x. $x = 0$ is always the best starting point since it gives an easy evaluation of y, in this case 0.5.

x	-4	0	6
y	-9.5	0.5	15.5

Figure 26 shows the graph marked with the linear law. Any pair of corresponding values of x and y on the graph will satisfy the law and therefore the original equation ($2y = 5x + 1$) from which it was derived.

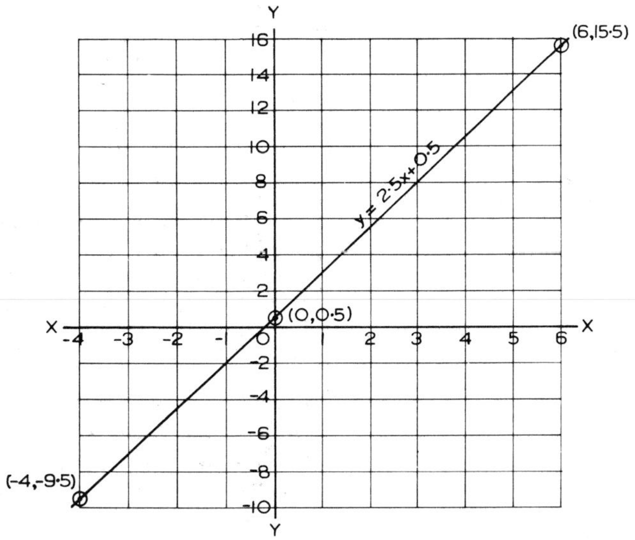

Figure 26

Example Draw the graph of $2y + 5x = 60$ for values of x from 0 to 20.

Rearrange the equation to a convenient form for the calculation of corresponding values.

$$2y + 5x = 60$$

Subtract $5x$ from both sides: $2y = -5x + 60$

Divide both sides by 2: $y = -2.5x + 30$

Three corresponding pairs:

x	0	10	20
y	30	5	−20

Note that, although x remains positive, y ranges from positive to negative and the first and fourth quadrants are required (Figure 27).

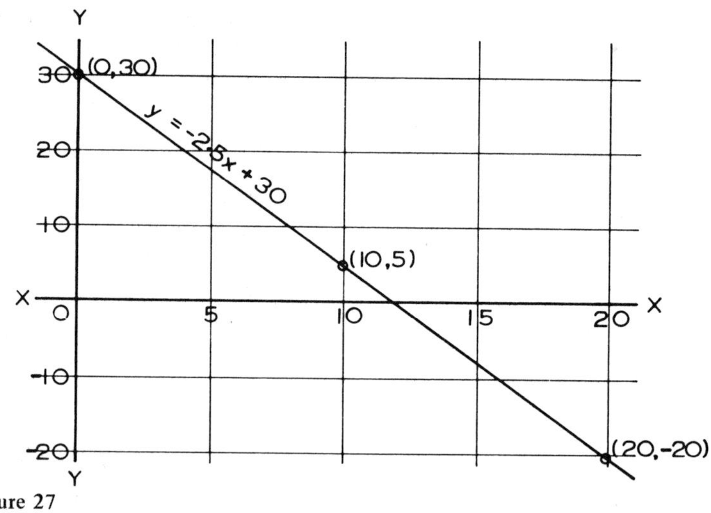

Figure 27

13.11 Reading graphs

Although in the previous examples only a limited number of points have been plotted, it is apparent that the coordinates of any point on the line give a pair of values of x and y that will satisfy the equation of the line. Thus it is a simple matter to match any given value of x with the corresponding value of y, and vice versa, from the graph itself.

Example Plot a graph of the relationship $\frac{1}{4}(3.2x + 2) = \frac{1}{10}(4y - 1)$. Use the graph to find values of y for all odd integral values of x between −4 and 8 and values of x for all even integral values of y between 9 and 17.

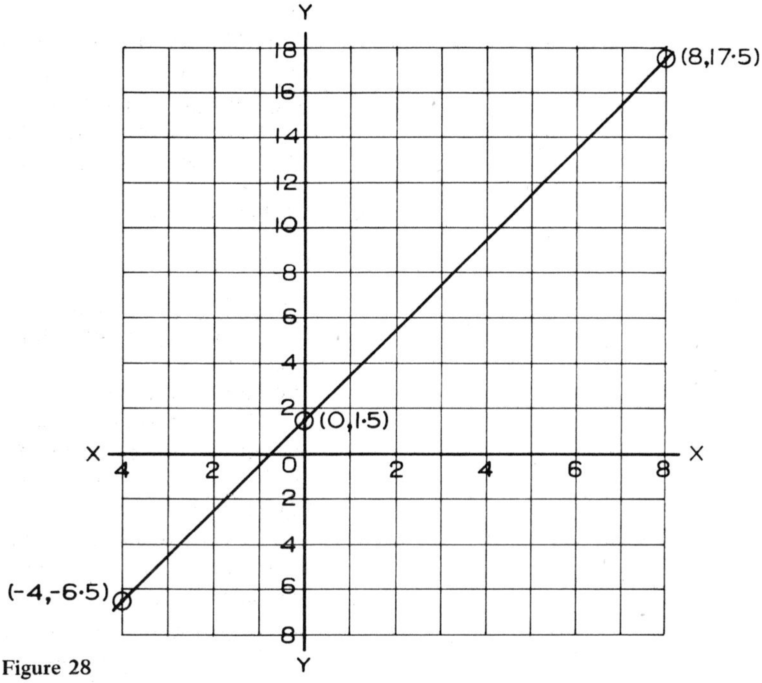

Figure 28

Rearrange the equation to find corresponding values.

	$\frac{1}{4}(3.2x + 2) = \frac{1}{10}(4y - 1)$
Multiply both side by 20:	$5(3.2x + 2) = 2(4y - 1)$
Expand brackets:	$16x + 10 = 8y - 2$
Add 2 to both sides and reverse:	$8y = 16x + 12$
Divide both sides by 8:	$y = 2x + 1.5$

Take three pairs of values:

x	−4	0	8
y	−6.5	1.5	17.5

From the graph (Figure 28) the required values of x and y are as tabulated below. For odd integral values of x:

x	−3	−1	1	3	5	7
y	*−4.5*	*0.5*	*3.5*	*7.5*	*11.5*	*15.5*

For even integral values of y:

y	10	12	14	16
x	*4.25*	*5.25*	*6.25*	*7.25*

In addition to reading off values from *within* the plotted range, it is often convenient to obtain values *beyond* that range. The following example illustrates this application.

Example The table shows corresponding values of head of water in metres and the resulting pressure in millibars. Plot a graph of the relationship and use it to estimate:
i) the pressure due to a 3 m head of water,
ii) the head of water required to produce a pressure of 500 mb.

Head of water (m)	0.5	1.5	2.5	3.5	4
Pressure (mb)	49	147	245	343	392

The highest pressure to be plotted is 500 mb and, from the fact that a pressure of 245 mb results from a head of 2.5 m, this would appear to correspond to a head of a little over 5 m. The scale is chosen to allow for these values.

It is conventional to plot the *independent variable* on the x axis and the *dependent variable* on the y axis. Since pressure is dependent on head it is plotted on the y axis. Results (from Figure 29):
i) Pressure due to a 3 m head of water = 294 mb. This is read off within the plotted range by *interpolation*, i.e. from between the plotted points.
ii) Head required to produce a pressure of 500 mb = 5.1 m. This result is read off from an extension of the line beyond the plotted points, a process termed *extrapolation*.

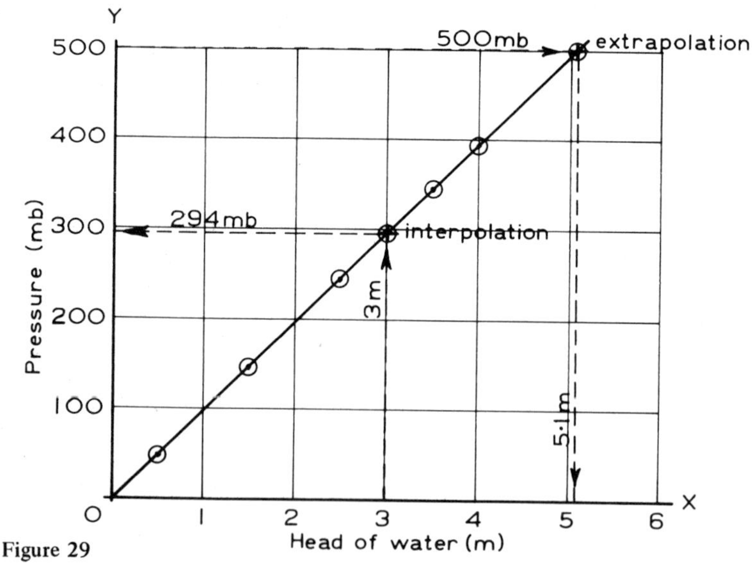

Figure 29

13.12 Gradient of a line

The *gradient* of a straight line is the ratio of the change in the value of y to the change in the value of x between any two points on the line.

In Figure 30 two points P and Q lie at (2, 3) and (7, 6) respectively on a straight-line graph. R is the point where the horizontal line through P meets the ordinate to Q. Then:

$$\text{Gradient of the line} = \frac{\text{change in } y}{\text{change in } x} = \frac{QR}{PR}$$

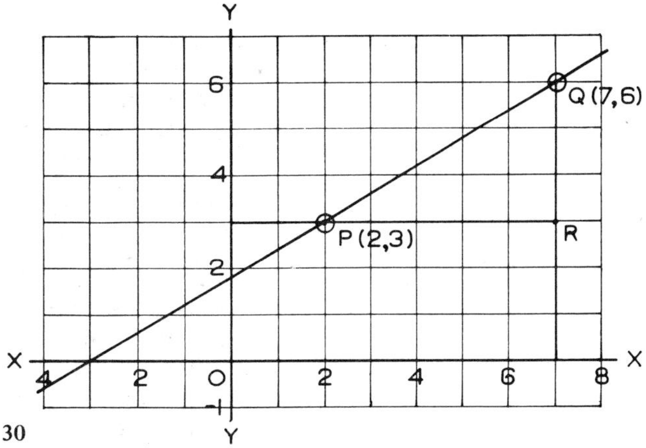

Figure 30

$$= \frac{6 - 3}{7 - 2}$$

$$= \tfrac{3}{5} = 0.6$$

Wherever P and Q are taken on the line the result will be the same.

The gradient of a straight line is important in so far as it represents the *rate of change of y relative to x* and therefore gives the value of the constant connecting the two variables. In the previous example, y is increasing at 0.6 of the rate of increase of x, and from any given starting point their relative values may be determined.

If a line has a gradient of 1 the line makes an angle of 45° with the x axis, since x and y are both increasing at the same rate. This is shown in Figure 31.

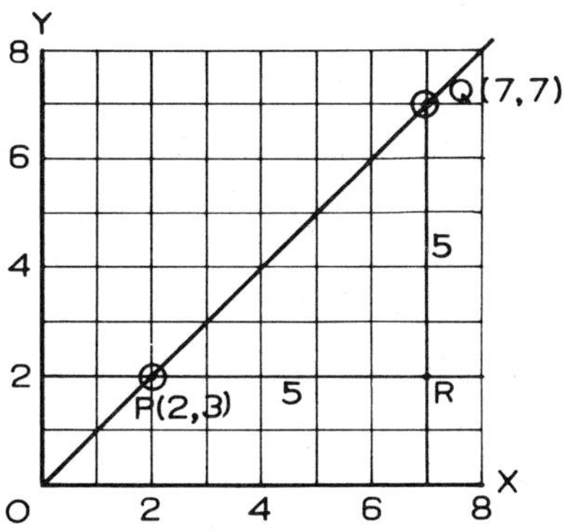

Figure 31

For any line steeper than 45° the gradient is greater than 1, since QR is greater than PR. Thus in Figure 32 the gradient is $\frac{6}{4} = 1.5$.

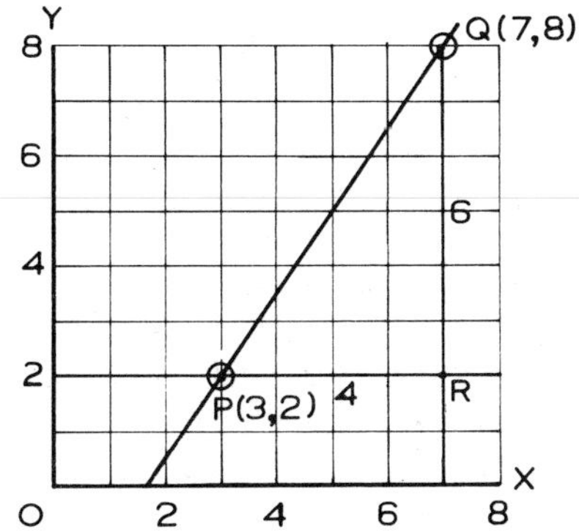

Figure 32

NEGATIVE GRADIENT

The gradient of a line is negative if there is a *negative change in y for a positive change in x*. In Figure 33 the change in the value of x between P and Q is positive, but the corresponding change in y is negative.

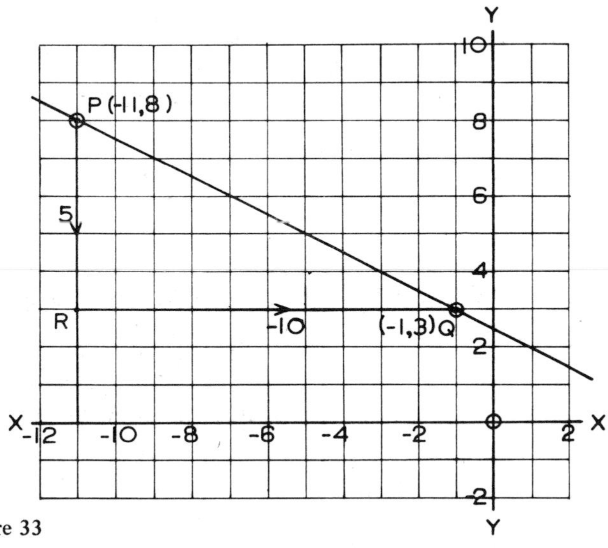

Figure 33

$$\text{Gradient} = \frac{8 - 3}{-11 - (-1)}$$

$$= \frac{5}{-10} = -0.5$$

In other words a line that rises from left to right has a positive gradient and a line that falls from left to right has a negative gradient.

Example A straight line has a gradient of 0.8 and passes through the point (−2, −5). Draw the graph and use it to complete the following pairs of coordinates of points that lie on the line: (3 y), (x, 1), (x, 5), (0.5, y).

The gradient 0.8 may be written as $\frac{8}{10}$, which indicates that y changes by 8 as x changes by 10. The slope is positive (i.e. upwards to the right).

The graph is shown in Figure 34. The point (−2, −5) is marked P, and a second point Q is plotted at a distance 8 above and 10 to the right of P. Join PQ, which now has a gradient of $\frac{8}{10}$ or 0.8. From the graph the required coordinates are (3, −1), (0.5, −2.5), (5.5, 1) and (10.5, 5).

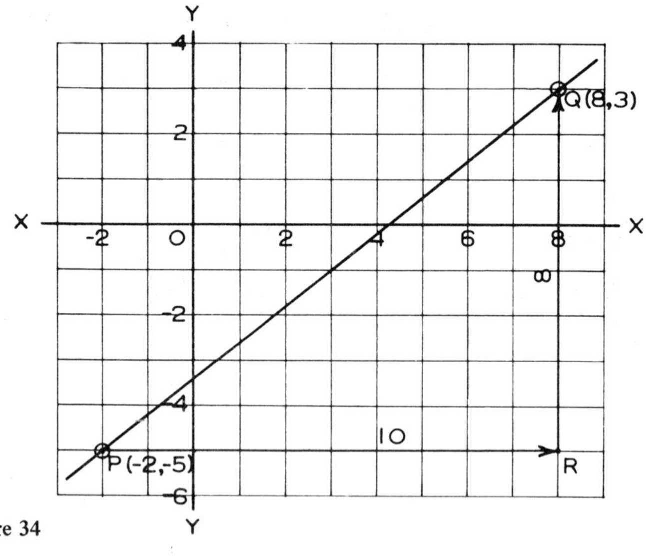

Figure 34

Exercise 13c

1. Plot the points (2, 3), (3, −1) and (−2, 1), and find the area of the triangle formed by joining them.

2. Plot the points (1.5, 0.5), (−2, 0.5), (−2, −1.5) and (1.5, −1.5), and determine the perimeter of the rectangle formed when they are joined.

3. Plot the graphs of the following equations for values of x between −3 and +3. In each case find the points at which the line intercepts the x and y axes.

a) $y = \frac{1}{2}x$ b) $y = x + 1$ c) $y = 3x + 4$ d) $y = 2x - 1$

4. Plot the graphs of the following linear equations, and name the figure that they enclose:

$y = 0.5x + 2$, $y = 3x - 5$, $y = 3x + 3.5$ and $y = 0.5x - 2.25$

5. Find the gradients of the graphs of the following equations:

a) $3y = 4x$ b) $\frac{1}{3}y = x - 1$ c) $2y - 5x = 4$ d) $\frac{x+1}{3} = \frac{y-2}{4}$

6. Draw the line that passes through the point $(2, -1.5)$ and has a gradient of 1.5. Where does it intercept the x and y axes?

7. Find the points where the line with a gradient of -0.75 and passing through the point $(-3, 4)$ intercepts the x and y axes.

SUMMARY OF D

Adjacent scales: two scales on a single line, zero marks coinciding.
Non-adjacent scales: scales on parallel lines reading in opposite directions the focal point at the intersection of lines joining corresponding values.
Rectangular axes: x and y axes at right angles.
Proportionality: direct dependence of one set of values upon another, producing a straight-line graph.
Layout, origin, signs, quadrants, coordinates (abscissa and ordinate): see Figures 35 and 36.
Scales: should be as large as possible in multiples of 2, 5 or 10.
Presentation: label axes clearly, do not crowd, use available space to the full.
Linear law: the equation connecting two variable quantities represented by a straight-line graph.
Independent variable: the key quantity, conventionally represented on the x axis.
Dependent variable: the quantity determined from the value of the independent variable, conventionally plotted on the y axis.
Interpolation: the reading of values from within the plotted range of a graph.
Extrapolation: the reading of values from the extension of a graph beyond the plotted range.
Gradient: ratio of change in y to change in x (Figures 37 and 38).

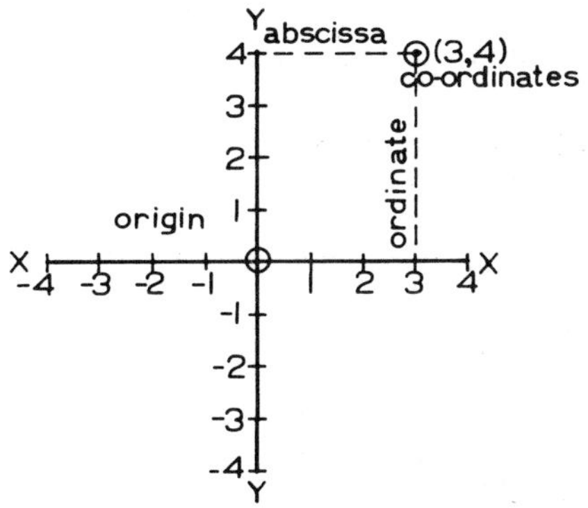

Figure 35

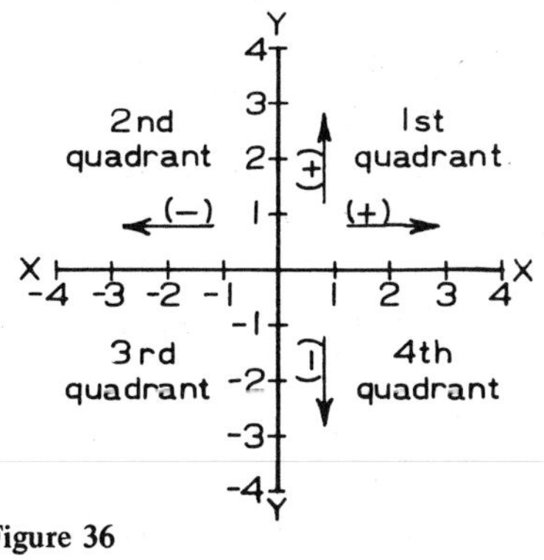

Figure 36

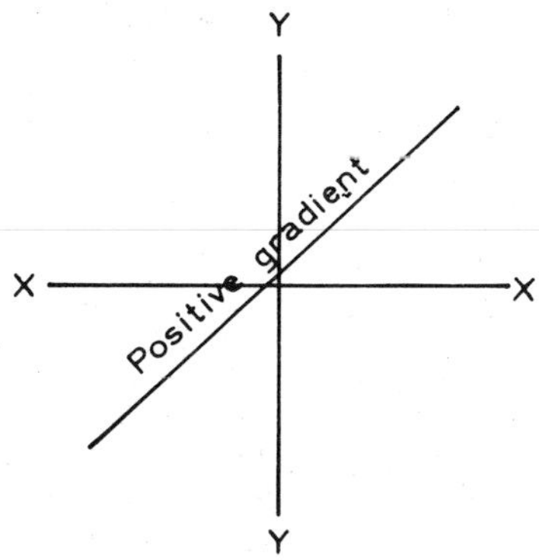

Figure 37

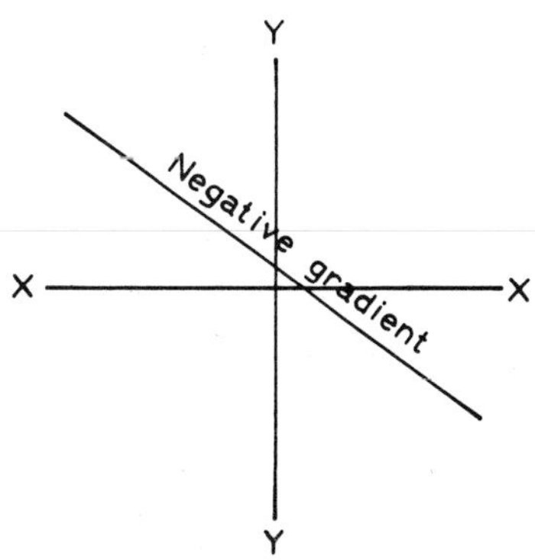

Figure 38

REVISION EXERCISE D

1. Construct adjacent scales to convert between the units of energy foot-pounds force and joules. 1 ft-lbf = 1.356 J. Base the scale on 10 units, and use it to convert 56 ft-1bf to joules and 339 J to foot-pound force.

2. On parallel lines construct non-adjacent scales to make conversions from horsepower to kilowatts. 1 hp = 745.7 W. From the scales find the equivalent of 35 hp in kilowatts and express 10 kW in terms of horse-power.

3. Wind pressure on exposed surfaces has the equivalent values of 100 $1bf/ft^2$ and 4.79 kN/m^2 in the imperial and SI systems. Construct a graph on rectangular axes to convert between the two systems. Use the graph to determine the metric equivalent of 75 lbf/ft^2 and the imperial equivalent of 8.5 kN/m^2.

4. In an experiment using a lifting machine, the efforts required to lift a range of loads were as follows:

Load (newtons)	90	180	270	360	450	540
Effort (newtons)	45	63	81	99	117	135

Plot a graph showing the relationship between load and effort and use it to estimate:
a) the probable load to be lifted by an effort of 40 N,
b) the effort likely to be required to lift a load of 150 N.

5. The lengths of a spring, measured while it supported a range of masses, are shown in the following table:

Mass supported (kg)	0.1	0.2	0.3	0.4	0.5	0.6
Spring length (mm)	165	172	179	186	193	200

Plot a graph of these values and use it to find:
a) the probable length of the spring before it was loaded,
b) the mass required to make the spring 175 mm long,
c) the mass required to increase the length of the spring by 25 mm.

6. The points $(-3, 7)$ and $(6, -2)$ are diagonally opposite vertices of a square. What are the coordinates of the square's other vertices?

7. Show that the points $(0, -1)$, $(-1, -4)$, $(2, 5)$ lie on a straight line, and that the points $(1, 0)$, $(-2, 1)$ and $(7, -2)$ lie on a second straight line perpendicular to the first.

8. Plot the points $(5, 4)$ and $(-3, -4)$, and find the coordinates of the mid-point of the line joining them.

9. Draw the straight line that passes through the points $(-4, 3)$ and $(5, -1)$, and find the points at which it intercepts the x and y axes.

10. Plot the graphs of the equations $2x = 14 - 7y$ and $7y = 8x - 21$, and find the coordinates of the point at which they intersect.

11. Plot the lines represented by the equations $y = 3x + 1$ and $y = 1 - 2x$, and find the gradient of each.

12. On one pair of axes, draw through the point $(2, 1)$ the lines that have the gradients 0.4, 2.2 and -1.5. In each case, find the intercepts on the x and y axes.

E Statistics

This chapter deals with:
Collection, tabulation and grouping of data
Tally diagrams; frequency; relative percentage frequency
Pie charts, 100% bar charts
Pictograms, vertical and horizontal bar charts
Histograms
Interpretation of statistical diagrams

14 COLLECTION AND PRESENTATION OF DATA

14.1 Surveys and samples

The collection of numerical data, and its tabulation and interpretation to show trends and predict future probabilities, is termed *statistical analysis*. The use of statistics in this way can be highly effective in increasing efficiency and productivity, ensuring consistent quality and improving safety. The gathering of the necessary information involves various activities such as counting numbers, measuring components, testing materials, extracting details from accounts and so on. Such processes are *statistical surveys*, and if the collected data is only a proportion of the whole range being investigated it is referred to as a *sample*. Samples obtained by counting numbers of items consist of *discrete* data. Surveys carried out by taking measurements produce samples of *continuous* data. Exercises in the collection of data are easily devised and may include such simple surveys as:
i) counting the number of cars or pedestrians using a certain entrance,
ii) measuring or weighing a specified number of bricks or other components from a stock,
iii) tabulating the results of laboratory experiments, such as absorption tests,
iv) collecting results of strength tests on materials.

14.2 Grouping data

After collection, the facts and figures need arranging and grouping in a way that will enable them to be read and understood quickly and completely. Some of the methods of presentation are shown in the following examples.

Example Absorption tests on twenty-five clay bricks selected from a batch of five hundred produced the following results. Arrange the data in groups of 1% range.

Percentage absorption by weight

15.5	17.4	15.1	18.4	15.9
16.2	16.3	18.2	15.3	16.5
17.3	18.4	15.2	17.5	16.7
15.7	17.5	16.5	16.2	17.3
18.1	17.7	17.8	18.4	17.4

These are values of the *variable*, and their grouping in ranges of 1% produces *classes*, the *class interval* being 1%.

This is continuous data, and the classes will be 15% to 15.9%, 16% to 16.9% and so on. The table below shows a convenient method of extracting the listed values into these class groups. Working methodically through the list, line by line, an oblique stroke (*tally* mark) is made for each item against its class description. After four tally marks the fifth member is represented by a horizontal line completing a set of five. The class total across each line is then easily determined.

% absorption class		*No. of bricks in class*
15 to 15.9	~~////~~ /	6
16 to 16.9	~~////~~ /	6
17 to 17.9	~~////~~ ///	8
18 to 18.9	~~////~~	5
	Total	25

14.3 Tally diagrams

The above method of tabulation forms a *tally diagram*. It may not always be appropriate to use this precise form of tallying, but similar methods are usually possible, as indicated in the next example.

Example A company engaged in the construction of high rise domestic buildings completed the following contracts in a certain year:

4 blocks each with 12 three-bedroom units,
12 two-bedroom units
and 10 one-bedroom units;
6 blocks each with 8 four-bedroom units,
4 three-bedroom units,
and 12 two-bedroom units;
4 blocks each with 14 two-bedroom units,
6 one-bedroom units,
and 10 bachelor flats (bedsitters).

Group the units according to the number of bedrooms but keeping the bedsitters separate. This data is *discrete*.

No. of bedrooms		*Units with this number*
4	(6 × 8)	48
3	(4 × 12) + (6 × 4)	72
2	(4 × 12) + (6 × 12) + (4 × 14)	176
1	(4 × 10) + (4 × 6)	64
Bedsitters	(4 × 10)	40
	Total	400

14.4 Frequency percentages

In each of the tables shown above, the right-hand column gives the *frequency* of each item or class, i.e. the number of times each occurs. When a frequency is expressed as a percentage of the total sample, the *relative percentage frequency* is obtained.

Two interesting and useful ways of expressing relative percentage frequencies are shown in the following diagrams. Taking the previous example and expressing the frequencies of the units as relative percentage frequencies, we have:

No. of bedrooms		*Relative % frequency*
4	$\frac{48}{400} \times 100\% =$	12%
3	$\frac{72}{400} \times 100\% =$	18%
2	$\frac{176}{400} \times 100\% =$	44%
1	$\frac{64}{400} \times 100\% =$	16%
Bedsitters	$\frac{40}{400} \times 100\% =$	10%

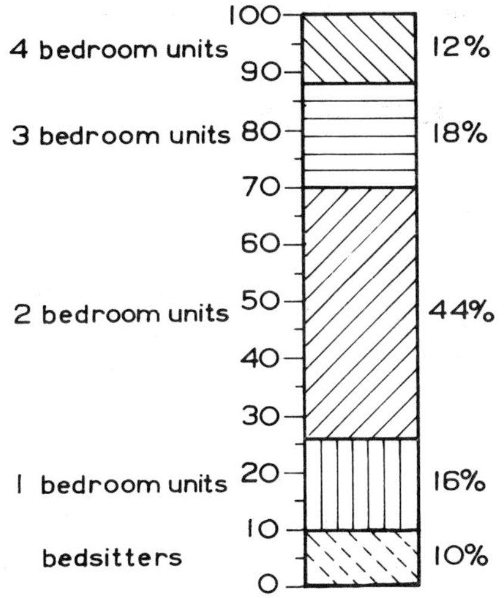

Figure 39

A parallel vertical column is drawn, its height representing the total number of items in the sample. The height is divided into parts in proportion to the frequencies. The relative percentage frequencies are thus represented by the areas of the blocks, which may be coloured or shaded for improved visual effect. This is a *100% bar chart* (Figure 39).

Another method of presentation is the *pie chart* (Figure 40). A circle is divided into sectors, the areas (and therefore the angles at the centre) of which are proportional to the frequencies. Construction, using a protractor, is simple once the angles have been calculated as the relative percentage frequencies of 360°.

No. of bedrooms		*Pie-chart angle*
4	$\frac{48}{400}$ of $360° =$	43.2°
3	$\frac{72}{400}$ of $360° =$	64.8°
2	$\frac{176}{400}$ of $360° =$	158.4°
1	$\frac{64}{400}$ of $360° =$	57.6°
Bedsitters	$\frac{40}{400}$ of $360° =$	36°

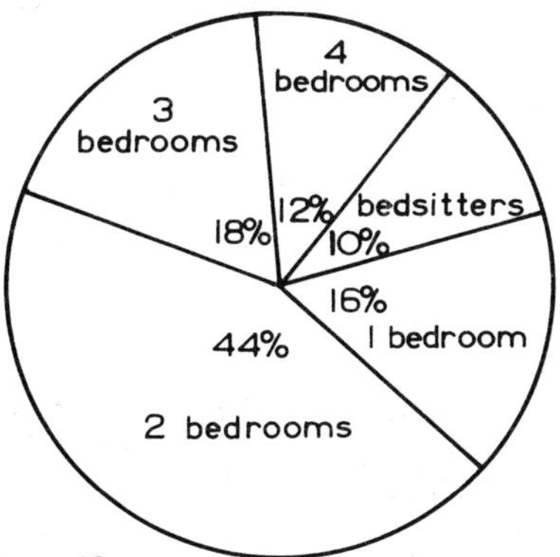

Figure 40

Construction to the nearest degree is adequate. As in the 100% bar chart, a better visual impression is produced by colouring or shading the sectors.

14.5 Frequency

We may now define 'frequency' as *the number of times a particular value or range of values occurs in a sample* and define *relative percentage frequency* as *frequency expressed as a percentage of the total number in the sample.*

14.6 Frequency diagrams

Another form of diagram used to present statistical information in an easily observed manner is the *pictogram*. This is best shown by an example.

Example The number of houses built in 1976 in a certain city was divided between three types of development body as follows:

Local authority building	150 houses
Private building	500 houses
Housing association building	50 houses

This information is depicted in the pictogram in Figure 41. Note that each symbol represents 100 houses and that the symbol is halved to show 50 houses. Varying the *size* of the symbol is not considered a suitable method of showing variation in numbers.

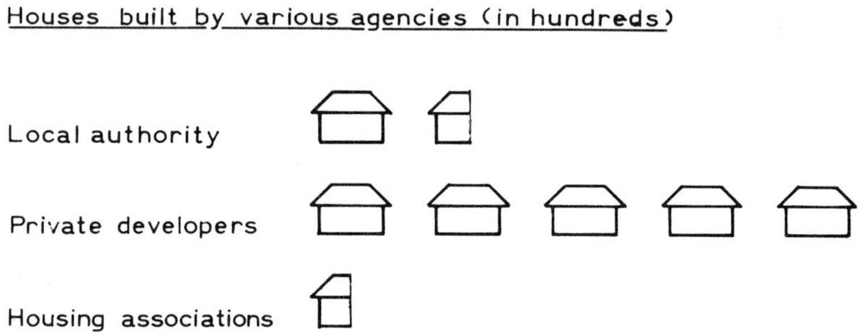

Figure 41

Yet another diagrammatic presentation is the Bar Chart, which may be horizontal or vertical as shown in the following example.

Example The value of work being carried out by a civil engineering company in a variety of locations is shown in the table below.

Nature of location	*Value of work in hand*
Urban	£2 400 000
Suburban	£3 250 000
Semi-rural	£750 000
Rural	£500 000
Isolated	£250 000

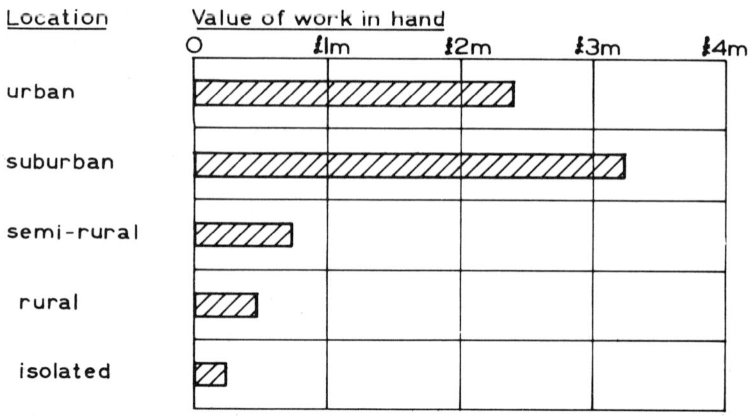

Figure 42

Figure 42 shows this distribution as a horizontal bar chart. The same information in a vertical bar chart is shown below in Figure 43. In both bar charts the lengths of the bars represent the values.

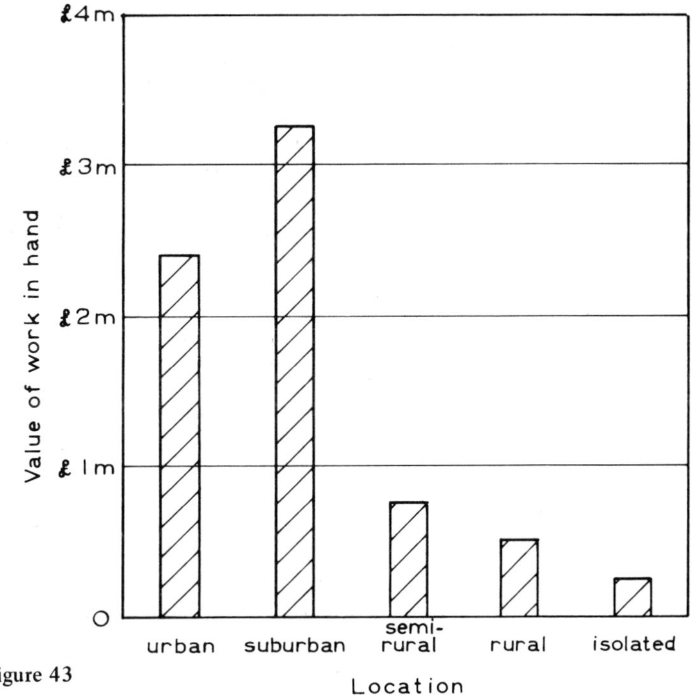

Figure 43

Exercise 14a

1. The following examination marks were obtained by eighty students in a science examination. Use a tally diagram to collect the results into classes, with a class interval of 10 marks.

67	5	63	11	66	32	64	32
72	74	59	43	34	24	45	69
68	41	49	72	42	51	42	52
21	57	59	47	83	14	77	58
51	31	43	73	92	78	35	18
23	86	8	89	26	81	49	57
46	15	56	33	53	48	76	62
55	47	48	44	71	54	29	25
35	65	80	22	65	31	51	61
48	37	60	95	59	66	91	50

2. A survey of fifty houses in a city centre showed the number of electrical power points per house to be as follows:

8	4	8	7	6	9	7	5	9	8
7	9	10	9	11	10	8	9	7	7
9	7	6	8	3	6	11	8	10	8
10	8	9	7	8	4	7	11	12	7
8	5	6	8	7	9	6	10	5	10

a) Is this data continuous or discrete?
b) Use a tally diagram to tabulate the results of the survey.

3. Compressive strength tests on fifty specimens of Scots pine in a timber research laboratory produced the following set of results:

Ultimate compressive stress in N/mm² of 50 Scots pine specimens tested parallel to grain

3.62	4.07	3.96	3.92	3.42	3.74	3.87	3.48	3.95	3.76
4.01	3.66	3.15	3.89	4.17	3.32	3.95	3.45	3.98	3.82
3.81	3.88	4.49	4.12	3.56	4.08	4.52	3.91	3.27	3.51
3.72	3.82	3.64	3.69	4.15	3.61	4.17	3.78	3.87	3.91
4.11	4.18	4.45	4.15	4.36	4.13	4.32	4.25	4.27	4.75

State whether the data is discrete or continuous, and use a tally diagram to collect it into classes having a class interval of 0.2 N/m² (i.e. 3.00 to 3.19, 3.20 to 3.39, etc.).

4. The following table shows the results of a survey of work completed by a group of building companies during a certain period in a particular district.

Type of building	*No. completed in period*
Bungalows	150
Maisonnettes	70
Semi-detached houses	255
Detached houses	145
Terraced houses	180

Produce a 100% bar chart to illustrate the relative percentage frequencies of the various types.

5. Construct a pie chart to illustrate the relative percentage frequencies of the faults listed in the following report.

Report of causes of persistent dampness from a survey of 150 houses

Nature of fault	*No. of faults reported*
Damp penetration through walls	14
Rising damp from floor level	40
Rain penetration at chimney stacks	24
Seepage of moisture at window sills	48
Saturation of ceilings from roof leaks	24

6. A large company employing 2 500 site workers has the following distribution of employees:

Nature of employment	*No. of employees*
Craftsmen	900
Labourers	1 250
Apprentices	200
Supervisors	100
Trainee supervisors	50

Devise a suitable pictogram to illustrate the distribution of personnel in the various types of employment.

7. Show by means of a vertical bar chart the following allocations of sales revenue for a company in a particular financial year:

Professional charges	£125 000
Wages	£512 500
Materials	£300 000
Overheads	£187 500
Taxation	£375 000

14.7 The histogram

A *histogram* is a diagram that shows *frequency distribution* at a glance. At first sight it would appear similar to a vertical bar chart, but it is in fact constructed on a different principle and has different applications.

Example A sample of 100 bricks taken from a production batch of 10 000 had the lengths shown below. Use a tally diagram to collect the lengths into groups in stages of 1 mm and construct a histogram of the frequency distribution.

Lengths of sample of 100 clay bricks in mm

215	214	215	217	212	216	211	213	212	218
213	215	216	215	215	217	215	212	217	216
215	212	215	216	211	218	213	215	216	214
213	216	214	215	217	214	216	214	215	213
217	216	215	213	214	215	215	216	214	215
216	215	216	215	215	216	215	215	218	216
214	217	214	215	216	215	217	214	214	213
216	213	215	217	215	216	213	215	216	219
214	215	216	214	215	215	217	214	214	215
215	214	215	215	215	214	216	215	217	214

Tally diagram

Length in mm		*No. of bricks of this length*
211	//	2
212	////	4
213	卌 ////	9
214	卌 卌 卌 ///	18
215	卌 卌 卌 卌 卌 卌 ////	34
216	卌 卌 卌 ////	19
217	卌 卌	10
218	///	3
219	/	1
	Total	100

Since it is not practical to measure bricks to a closer accuracy than 1 mm, it is assumed that those classified as 217 mm lie between 216.5 mm and 217.5 mm, and so on, and the widths of the blocks forming the histogram (Figure 44) are based on the stated lengths as central values.

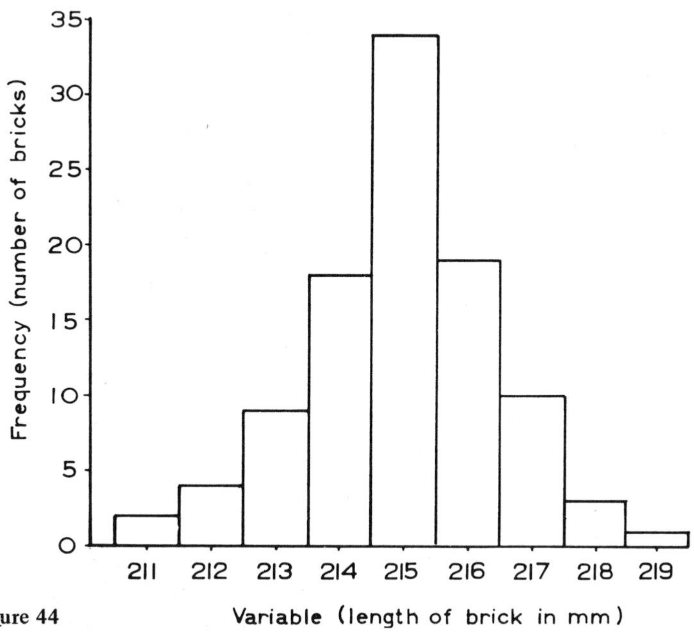

Figure 44

Example Thirty students were each given a specimen of elm and asked to weigh and measure it carefully and then to calculate its density. The following results were obtained and their grouping in a *tally diagram* is shown, followed by a *histogram* of the frequency distribution.

Densities of 30 specimens of elm in kg/m^3

568	558	577	563	574	571
557	572	587	592	556	599
571	586	547	535	566	584
579	549	582	574	542	562
562	567	591	553	581	559

The results are grouped in classes with a class interval of 10 kg/m^3 in the following tally diagram.

Density range in kg/m^3		*No. of specimens in range*
535 to 544	//	2
545 to 554	///	3
555 to 564	~~////~~ //	7
565 to 574	~~////~~ ///	8
575 to 584	~~////~~	5
585 to 594	////	4
595 to 604	/	1
	Total	30

Note that, since the variable values are given to the nearest integer, the gap of 1 unit between classes is acceptable. Had the densities been given to the nearest 0.1 kg/m^3 the classes would have appeared as 535 to 544.9, 545 to 554.9, etc.

The histogram is now plotted on a horizontal base line along which the density classes are represented by adjacent spaces. The frequency values are plotted vertically forming rectangular blocks the *areas* of which represent the frequencies (Figure 45).

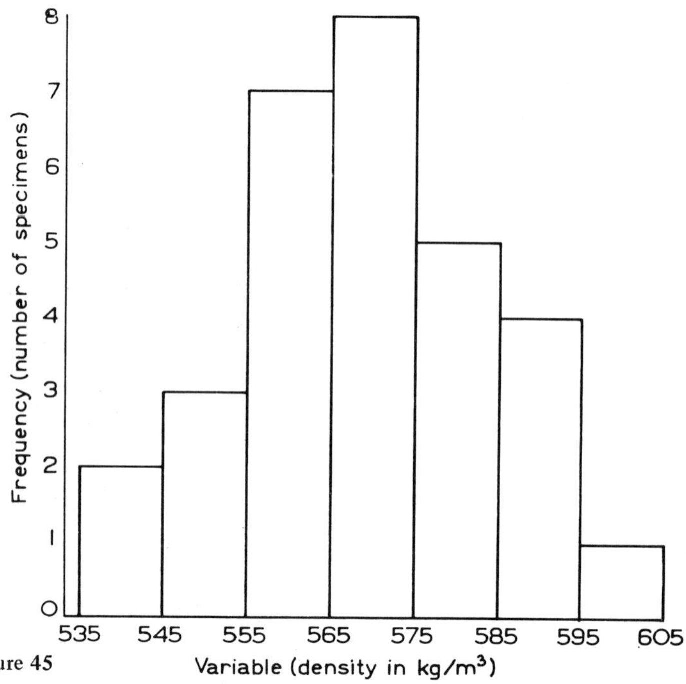

Figure 45

14.8 Clarity of presentation

Look back through the examples used in this chapter and note the importance of clear and accurate labelling of diagrams.

- The units of listed values should be stated clearly.
- Tally diagrams should have fully descriptive headings.
- 100% bar charts and pie charts should be constructed accurately and labelled clearly with descriptions and values of their parts.
- The horizontal and vertical axes of bar charts should be marked with the quantities represented and the units used. Scaled divisions should be clear.
- Pictograms should be drawn neatly and the symbols should be of equal size, their number representing the quantity. An indication of the symbol's value must be given.
- Histograms should be carefully and accurately constructed, the width of the blocks spanning the range represented and the vertical scale marked clearly on the axis.

Note that whereas a bar chart represents values by the *lengths* of the bars, a histogram represents frequencies by the *areas* of the blocks.

14.9 Interpretation of statistical diagrams

Some examples of the diagrams dealt with so far are given below, with written interpretations of the data they represent.

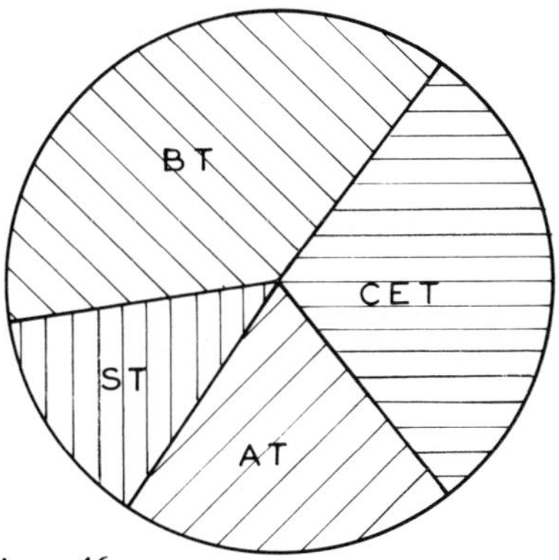

Figure 46

Example Figure 46 is a pie chart showing the distribution of 120 students in a TEC course (BT = building technicians, CET = civil engineering technicians, AT = architectural technicians, ST = surveying technicians). Measurement of the angles gives the numbers of students in the various groups:

Group	*Angle*		*No. of technicians*
Building	135°	$\frac{135}{360} \times 120 =$	45
Civils	105°	$\frac{105}{360} \times 120 =$	35
Architectural	72°	$\frac{72}{360} \times 120 =$	24
Surveying	48°	$\frac{48}{360} \times 120 =$	16
		Total	120

The chart shows that, of the 120 students in the course, more than a third are building technicians, a little over a quarter are civil engineering technicians, just over a sixth are architectural technicians and approximately an eighth are surveying technicians. If required the percentages may be calculated as follows (these are to the nearest 0.5%):

Building	$\frac{135}{360} \times 100\% = 37.5\%$
Civils	$\frac{105}{360} \times 100\% = 29\%$
Architectural	$\frac{72}{360} \times 100\% = 20\%$
Surveying	$\frac{48}{360} \times 100\% = 13.5\%$

Example For Electrical Installations Ltd, the value of work completed in a 5 year period (in £10 000) was as shown in Figure 47.

This *pictogram* shows an increase in trading of the company from £30 000 in 1972 to £55 000 in 1974, followed by a decrease in the following year to £40 000 and a fall to below the 1971–72 figure in 1976 to £25 000.

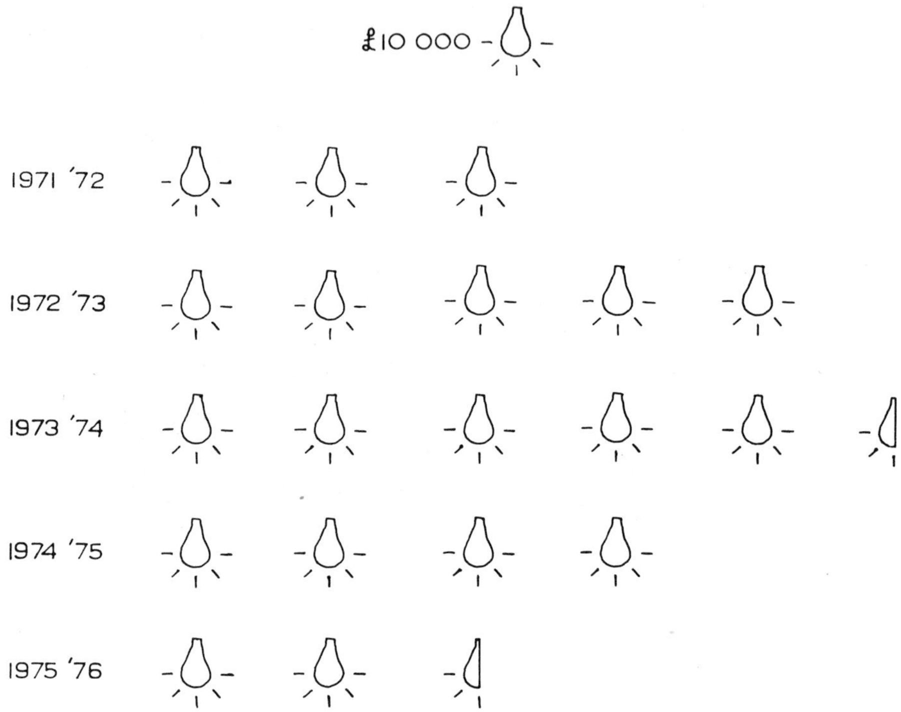

Figure 47

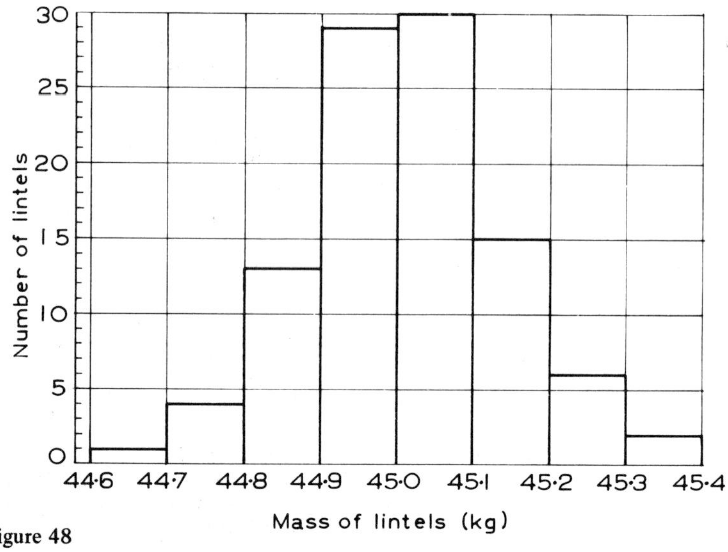

Figure 48

Example Figure 48 is a histogram showing frequencies of mass values of 100 precast concrete lintels designed to weigh 45 kg.

The histogram shows the frequency distribution to be reasonably well balanced about the design value of 45 kg, but with fewer underweight than overweight specimens. More than 60% of the sample lies in a range of 0.1 kg on either side of 45 kg, and over 90% within a range of 0.2 kg on either side of 45 kg.

Exercise 14b

1. Construct a histogram of the data given in question 1, Exercise 14a.

2. An analysis of the weekly wage bill of a firm with a hundred employees produced the following information. Present this data in the form of a histogram.

Weekly wage	*No. earning this wage*
£20 to £30	5
£30 to £40	18
£40 to £50	27
£50 to £60	24
£60 to £70	13
£70 to £80	8
£80 to £90	5

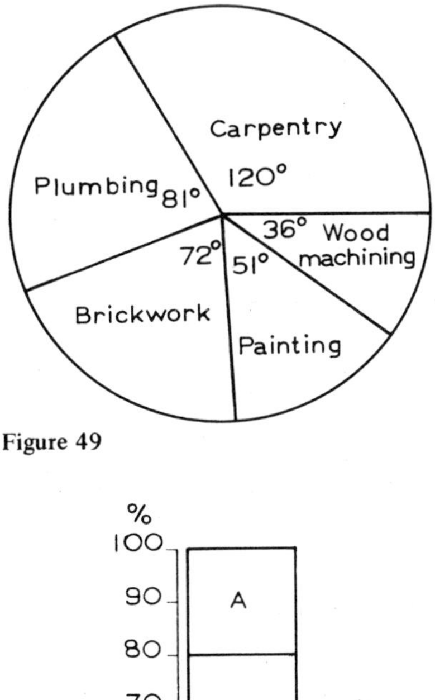

Figure 49

3. The compressive strengths of sixty concrete test cubes are listed below. Collect these values into nine suitable classes and draw a histogram of the frequency distribution.

Compressive strength in N/mm²

14.5	26.5	20.5	29.5	33.5	18.0	30.5	28.0	22.5	35.0
32.5	33.0	17.5	23.5	24.0	30.0	24.5	28.5	34.0	33.5
27.0	25.5	26.5	27.0	38.5	25.0	21.0	32.0	25.5	35.5
26.0	29.5	22.0	30.0	27.5	31.5	29.0	41.0	31.5	38.0
32.5	28.5	25.0	23.0	31.0	22.5	28.0	34.5	27.5	36.5
21.5	29.0	23.5	24.0	18.5	30.5	27.0	24.5	31.0	37.5

Give a concise summary of the information contained in each of the diagrams in questions 4 to 8.

4. Pie chart showing the crafts of 240 apprentices (Figure 49).

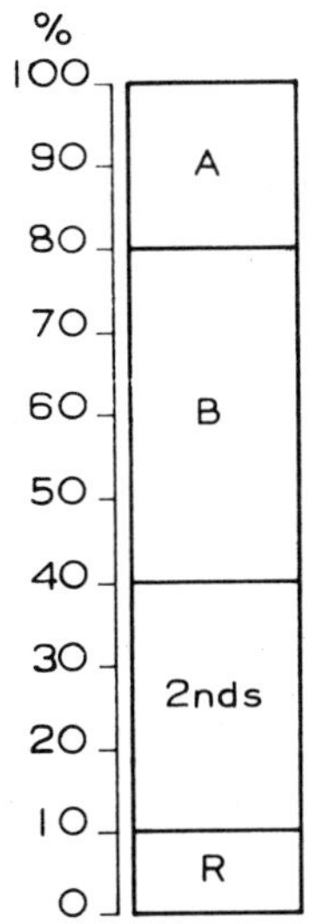

Figure 50

5. 100% bar chart showing the result of sorting 10 000 bricks from a clamp (Figure 50).

A Class A facings

B Class B facings

2nds Second-class stocks

R Rejects

6. Bar chart showing the measurements of selected trees in a plantation (Figure 51).

7. Cars serviced during the first three months of 1977. MMMM represents 100 cars.

Garage	*Cars serviced*
Alpha (London)	MMMM MMMM MMMM MMMM MMMM MMMM MMMM
Beta (Brighton)	MMMM MMMM MMMM MMMM MMMM MM
Gamma (Guildford)	MMMM MMMM MMMM MMM
Delta (Petworth)	MMMM M

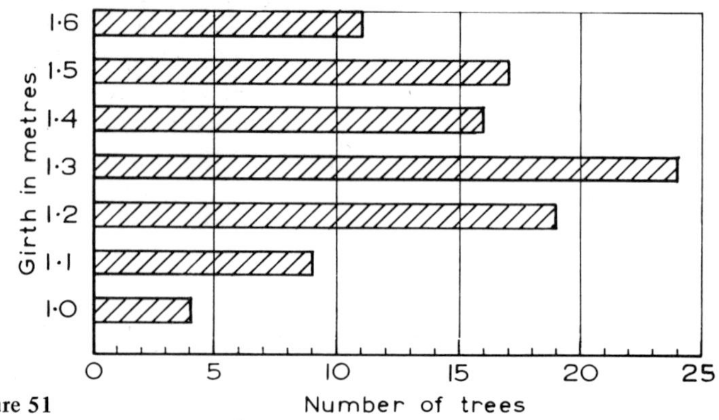

Figure 51

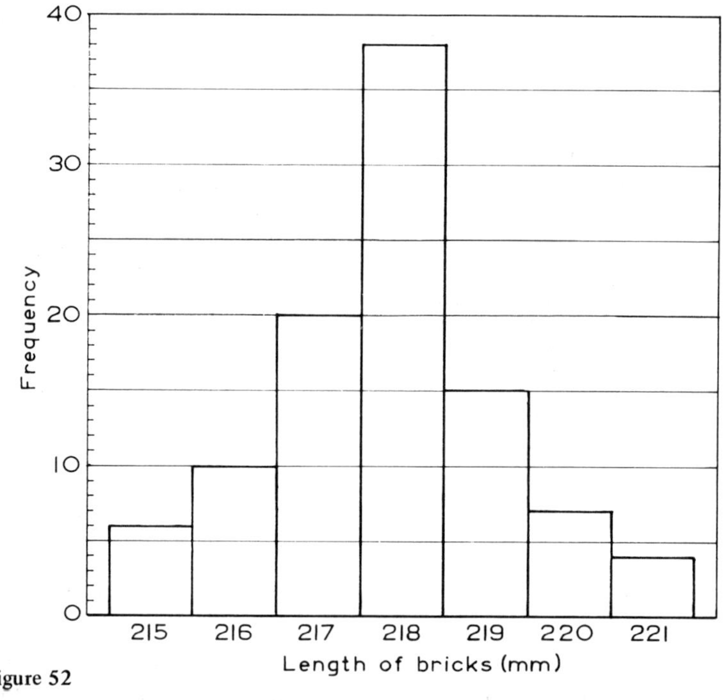

Figure 52

8. Histogram showing the measured lengths of a sample of bricks from a kiln batch (Figure 52).

SUMMARY OF E

Statistical survey: the collection of data.
Sample: representative facts and figures collected from a larger range of information.
Continuous data: data obtained by taking measurements.
Discrete data: data obtained by counting numbers.
Tallying: marking off items in groups of five, as ~~||||~~.
Frequency: the number of times a given value, or class of values, occurs in a sample.
Class: a group or range.
Relative percentage frequency: a frequency expressed as a percentage of the total number in the sample.

Pictogram: symbolic representation in pictures: the symbol should not vary in size but only in quantity. The value represented by the symbol must be stated.
Histogram: a frequency distribution diagram in adjacent vertical bars.
Presentation: clarity and full descriptions are essential. Units must be indicated.

REVISION EXERCISE E

1. The following list of marks resulted from a mathematics examination taken by a group of students. Use a tally diagram to group the marks into classes having a class interval of 10 marks, and draw a histogram of the frequency distribution.

25	82	12	63	57	51	35	44
59	45	44	58	3	60	52	61
25	40	63	53	27	50	53	90
81	58	32	19	72	37	30	62
48	62	38	37	51	75	65	47

2. The specialist craftsmen exployed by a contractor are listed below. Present the information in the form of (a) a pie chart and (b) a 100% bar chart.

Carpenters	50	Plasterers	5
Bricklayers	35	Painters	15
Plumbers	20		

3. Construct a pictogram of the following data, using the symbol 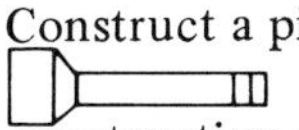 to represent each 100 m of pipeline. Length of pipeline construction completed in an industrial drainage scheme during the first six months of 1977:

Month	*Pipeline completed*
January	50 m
February	150 m
March	300 m
April	450 m
May	600 m
June	250 m

4. Each pound of a company's expenditure on machinery (excluding initial cost and depreciation) is made up as follows: maintenance 25p, fuel 45p, spare parts (stock) 20p, insurance 10p. Show this allocation in a pie chart.

5. The useful life of a component in an electrically operated hoist is indicated in the following results of tests on 200 such components. Construct a histogram of the frequency distribution.

Useful life in hours	*Frequency*
1 000 to 1 049	5
1 050 to 1 099	18
1 100 to 1 149	28
1 150 to 1 199	86
1 200 to 1 249	36
1 250 to 1 299	20
1 300 to 1 350	7

6. The output of a commercial vehicle coachbuilding firm for a five-year period is shown below. Use a pictogram to illustrate the data.

Year	*Output of vehicles*
1971	4 500
1972	5 750
1973	6 250
1974	6 000
1975	5 000

7. A survey of the number of vehicles using a certain road junction during the period 8 a.m. to 8 p.m. produced the following results. Show the data on a vertical bar chart.

Day of week	Mon	Tue	Wed	Thu	Fri	Sat	Sun
Vehicles	420	550	350	580	750	890	300

8. Strength tests of eighty specimens of an alloy produced the following results. Use a tally diagram to collect the data into five classess and show the frequency distribution by means of a) a histogram, b) a horizontal bar chart.

Ultimate stress in N/mm²

345	315	349	333	338	344	349	355	372	366
327	341	342	347	336	356	327	325	346	365
383	389	387	395	392	363	373	331	342	328
347	331	374	369	335	351	338	354	345	327
357	365	362	336	368	325	351	310	359	375
322	379	361	332	343	321	358	367	370	378
371	362	363	364	376	329	377	330	352	353
305	341	322	350	327	343	331	348	331	321

F Geometry

This chapter deals with:
Lines and angles; parallel lines
Triangles, types and properties; complementary angles
Theorem of Pythagoras; construction of triangles
Properties of the circle; sector and segment, radians
Linear properties and areas of quadrilaterals
Area of the circle
Volumes of solids: prism, cylinder, pyramid, cone and sphere
Surface areas of prisms and cylinders
Composite solids

15 LINES AND ANGLES

15.1 Rotation

Rotational displacement is measured in revolutions. Each revolution is divided into 360 divisions, each of which is one degree (1°). Although 1° appears to be a very small measure of rotation, the distance it represents becomes greater as the distance from the centre of revolution increases (Figure 53).

For accuracy in surveying and setting out and in navigation, the degree is again divided into sixty parts, each of which is one minute (1′). For extreme accuracy over long distances, the minute is divided into sixty parts, each of which is one second (1″).

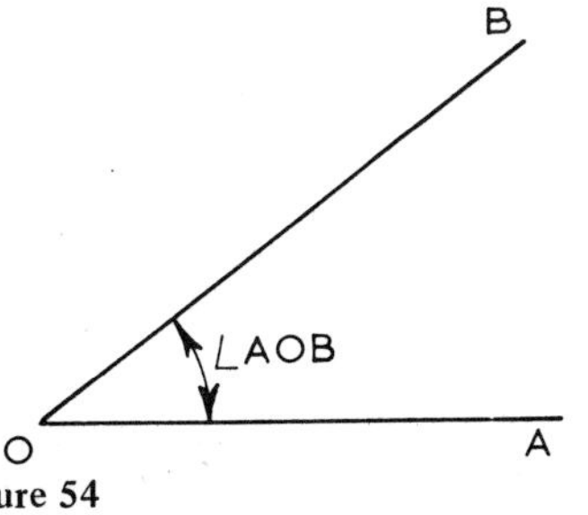

Figure 54

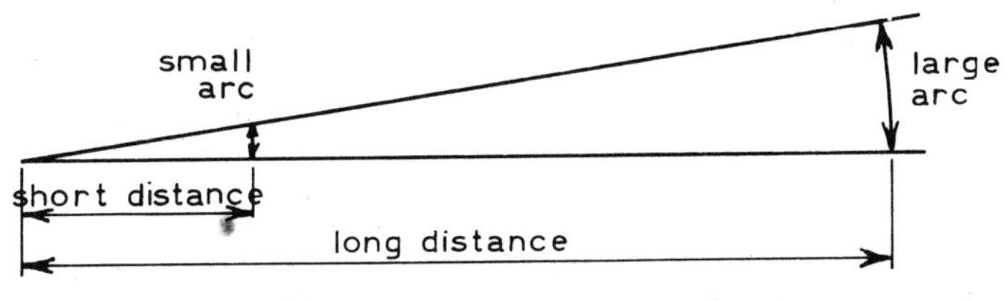

Figure 53

15.2 Angles

When two straight lines meet at a point they form an *angle*. In Figure 54, if the line OA revolves about O to a new position OB the angle AOB is formed. The rotational displacement from OA to OB is the size of the angle and may be measured in degrees. The point O is the *vertex* of the angle, and OA and OB are the *arms* of the angle.

An angle formed by one quarter of a revolution is a *right angle* and contains 90° as shown in Figure 55. OA and OB are *perpendicular* to one another.

Angles less than 90° are *acute* angles, and angles between 90° and 180° are *obtuse* angles. An angle of 180° (two right angles) forms a straight line and is sometimes referred to as a *straight angle*. Angles greater than 180° are *reflex angles*. Examples of angles are shown in Figure 56 overleaf.

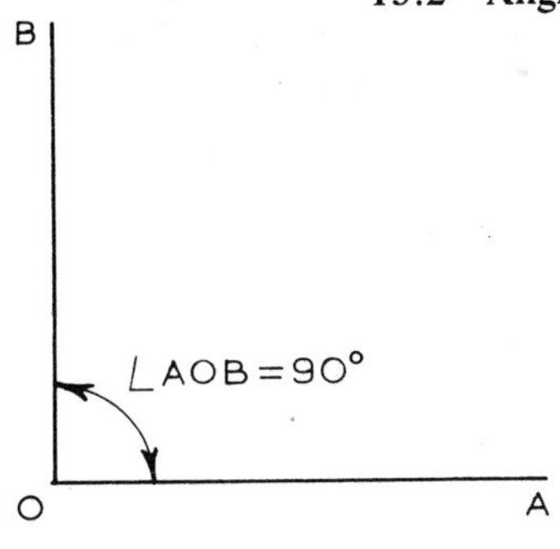

Figure 55

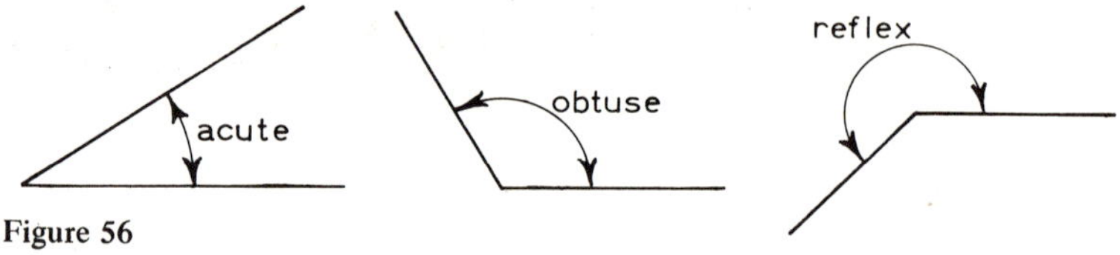

Figure 56

15.3 Vertically opposite angles

When two straight lines intersect, vertically opposite angles are formed. In Figure 57 ∠AOC and ∠BOD are vertically opposite, as also are ∠AOD and ∠BOC.

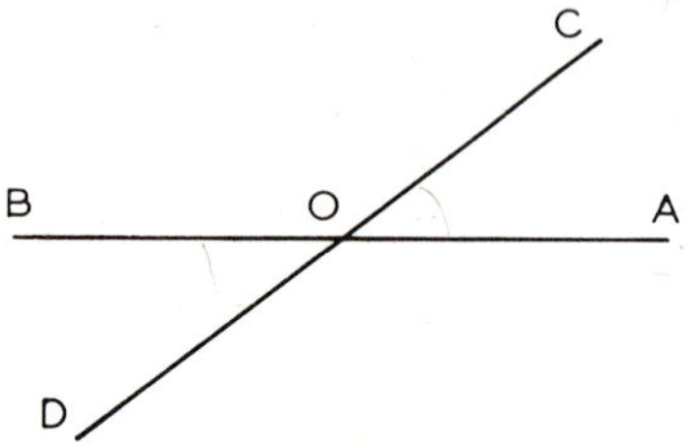

Figure 57

On AB ∠AOC = 180° – ∠BOC and

on CD ∠BOD = 180° – ∠BOC

∴ ∠AOC = ∠BOC

i.e. *vertically opposite angles are equal.*

15.4 Parallel lines

Parallel straight lines are lines in the same plane that never meet, no matter how far they are produced. Parallel lines are conventionally depicted by the use of arrows, as ⇉ . Figure 58 shows a pair of parallel lines intersected by a *transverse* line, forming eight angles. The angles 1, 2, 7 and 8 are *exterior* angles, and angles 3, 4, 5 and 6 are *interior* angles.

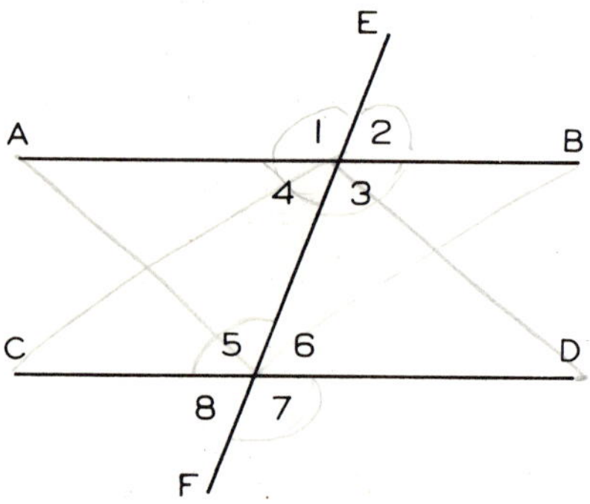

Figure 58

The pairs of angles (1, 3), (5, 7), (2, 4) and (6, 8) are vertically opposite angles.

The angles 3 and 5 are *alternate* angles, as are the angles 4 and 6. *Alternate angles are equal.*

The exterior angle 1 and its *interior opposite* angle 5 form a pair of *corresponding angles*; (2, 6), (4, 8) and (3, 7) are also pairs of corresponding angles. *Corresponding angles are equal.*

Angles such as 1 and 2, which together form two right angles, are *supplementary angles.*

15.5 Use of angle relationships

The angle properties referred to in section 15.4 may be used to identify parallel lines.

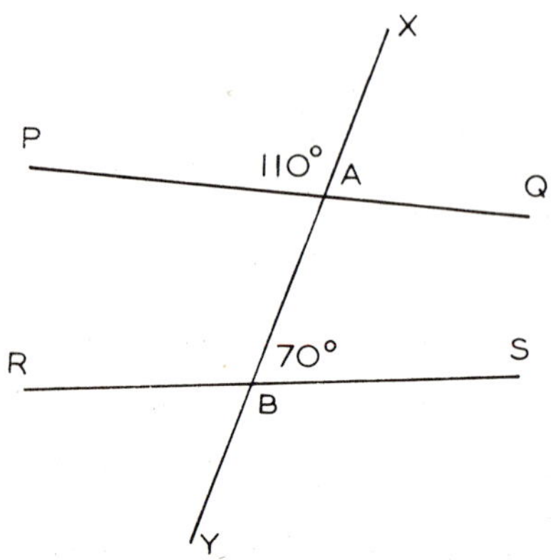

Figure 59

Example Two straight lines PQ and RS are cut by a third line XY, such that PQ is intercepted at A and RS at B. If ∠PAX = 110° and ∠SBA = 70°, show that PQ and RS are parallel.

Note that Figure 59 is drawn without assuming that the lines are parallel.

∠PAX + ∠QAX = 180° (supplementary angles)

∴ ∠QAX = 180° – 110° = 70°

∴ ∠QAX = ∠SBA

These are therefore corresponding angles, and PQ and RS are parallel.

Exercise 15 Find the unknown angles in the following figures:

1.

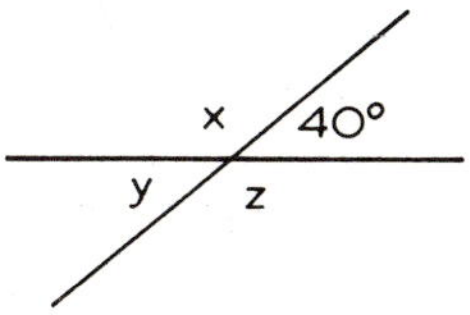

Figure 60

2.

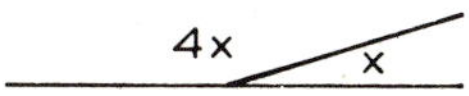

Figure 61

3.

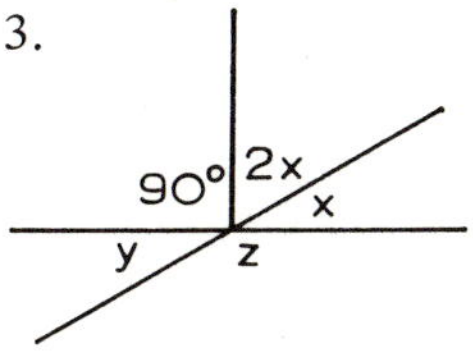

Figure 62

4.

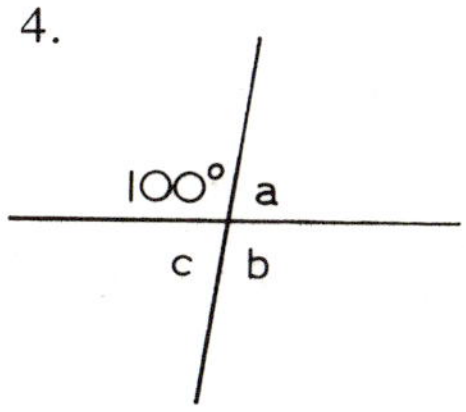

Figure 63

5.

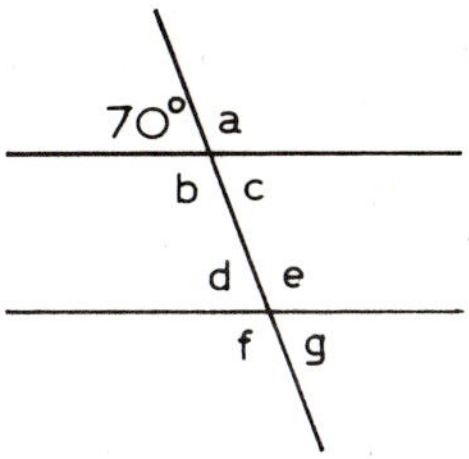

Figure 64

6.

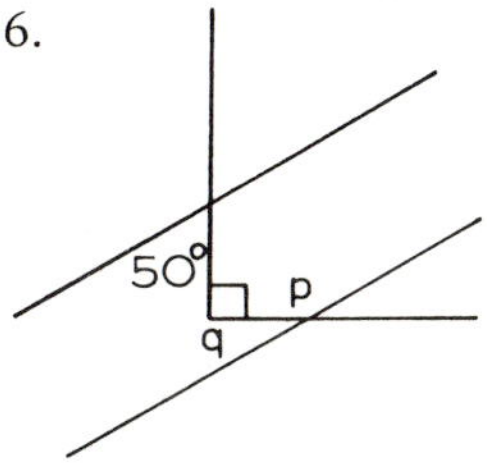

Figure 65

7.

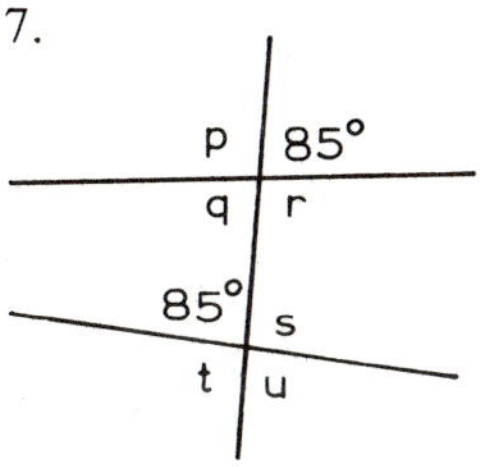

Figure 66

8.

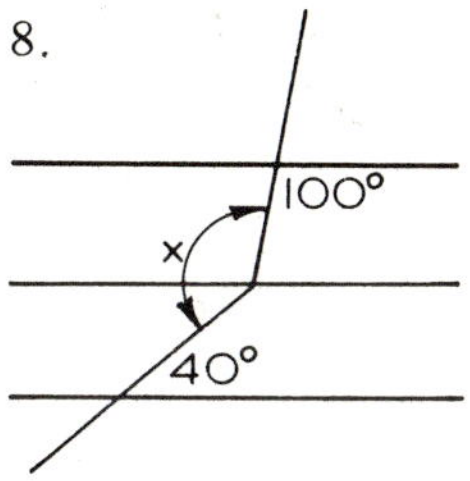

Figure 67

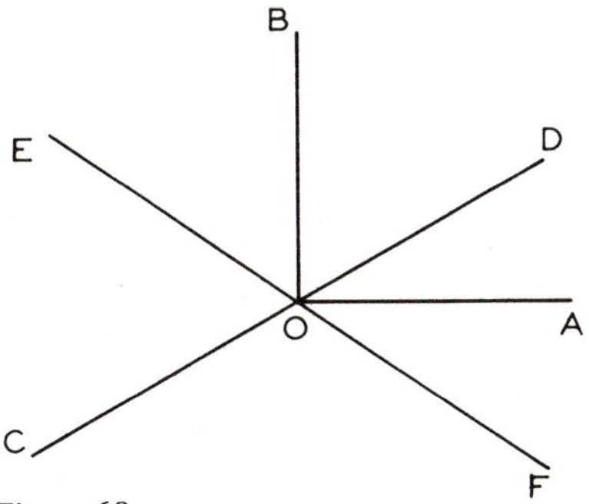

Figure 68

9. Find the supplements of 127°, 46° 30′ and 98° 55′.

10. In Figure 68 OA and OB are perpendicular straight lines, and CD and EF are straight lines intersecting at O; ∠AOD = 30°. Describe the angles AOD, AOB and AOC and find the values of the angles BOD and BOC.

11. What is the greatest number of obtuse angles that may be contained in one revolution?

12. What is the least number of acute angles that may be contained in three right angles?

16 TRIANGLES

16.1 Angles in a triangle

A triangle is a plane figure bounded by three straight lines, and therefore has three interior angles. The sum of the interior angles in any triangle is equal to two right angles (180°).

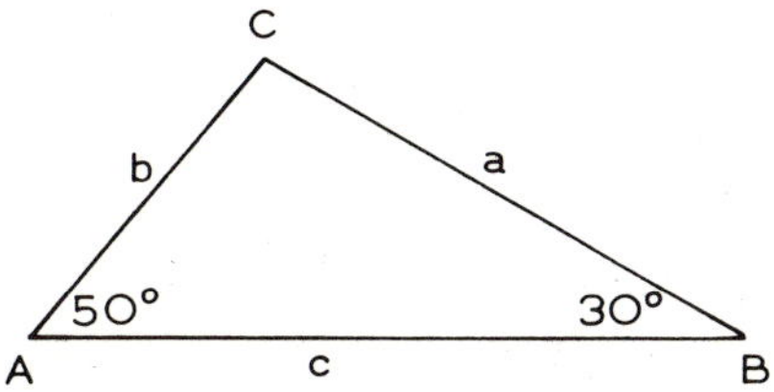

Figure 69

In Figure 69: $\angle A + \angle B + \angle C = 180°$

$\therefore \angle C = 180° - (\angle A + \angle B)$

$= 180° - (50° + 30°)$

$= \underline{100°}$

If one side of a triangle is produced, the angle formed with the adjacent side is called an *exterior* angle (∠ACD in Figure 70).

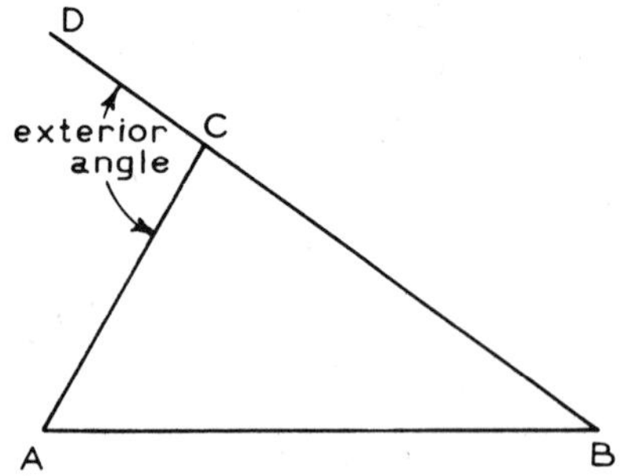

Figure 70

Since $\angle ABC + \angle BAC = 180° - \angle ACB$ (angles in a triangle)

and $\angle ACD = 180° - \angle ACB$ (supplementary angles)

$\therefore \angle ACD = \angle ABC + \angle BAC$

i.e. *the exterior angle is equal to the sum of the two interior opposite angles.*

In any triangle the longest side is opposite the greatest angle, and the shortest side is opposite the smallest angle.

It is conventional to label triangles with capital letters at the vertices and the corresponding lower- case letters on the respective opposite sides: i.e. *a* is the side opposite the vertex A, and so on.

16.2 Types of triangle

Triangles may be classified by the relationships between their sides or by the relationships between their angles (Figure 71).

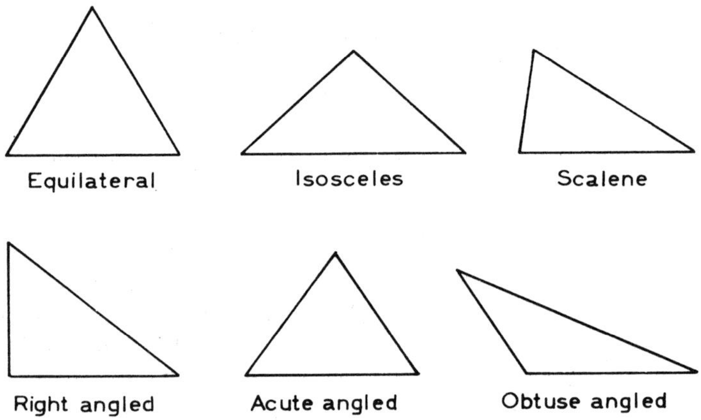

Figure 71

Equilateral triangle: all sides equal
Isosceles triangle: two sides equal
Scalene triangle: all sides of different length
Right-angled triangle: one angle of 90°
Acute-angled triangle: *all* angles acute
Obtuse-angled triangle: *one* angle obtuse

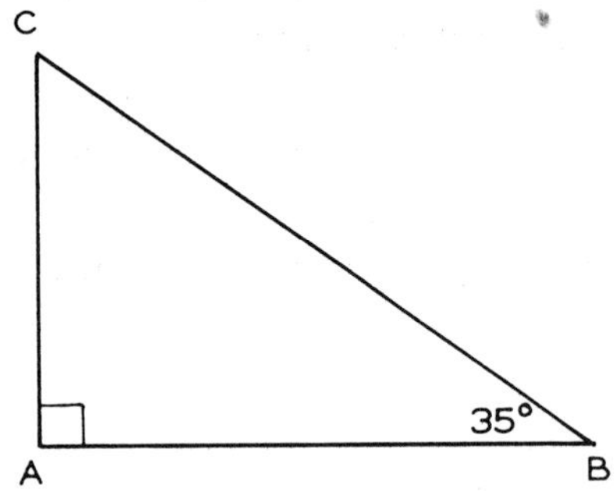

Figure 72

16.3 Complementary angles

In a right angled triangle the two acute angles have the sum of 90°. Such angles are *complementary angles*, together forming one right angle; e.g. 30° and 60° are complementary angles.

In Figure 72, $\angle C$ is the complement of $\angle B$

$\therefore \angle C = 90° - 35° = \underline{55°}$

Exercise 16a

1. Two angles of a triangle are 35° and 115°. Name the type of triangle and find the third angle.

2. What are the other two angles in a right-angled isosceles triangle?

3. The angle opposite the longest side, *a*, of a triangle ABC is 82° and $\angle B$ is 53°. Which is the shortest side of the triangle and what is the value of the third angle?

4. Sketch a triangle ABC, and draw a line through the vertex and parallel to the base. Use the diagram to show that the sum of the angles in a triangle is equal to two right angles.

5. The side BC of a triangle ABC is produced to D, and the exterior angle ACD is 127°. ∠BAC = 51°. Find the two remaining interior angles.

6. Find the complementary angles to each of the following: 17° 30′, 83° 10′, 27° and 52° 45′.

7. Find the supplement to each of the following angles: 83° 10′, 172° 30′, 150° and 25° 50′.

16.4 The theorem of Pythagoras

This theorem states that *the square on the hypotenuse of a right angled triangle is equal to the sum of the squares on the other two sides.* This relates to the *areas* of the squares on the sides and is illustrated in Figure 73.

If two sides of a right-angled triangle are known, the third may be calculated using the theorem.

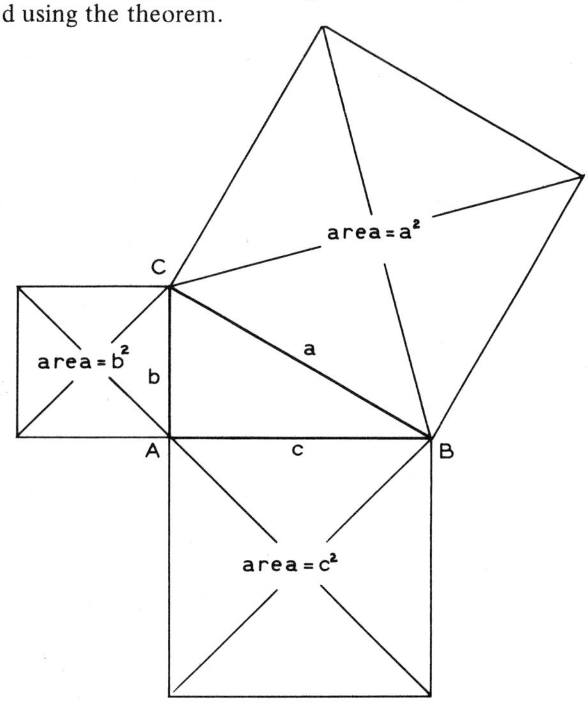

Figure 73 $a^2 = b^2 + c^2$

Example In a right-angled triangle (Figure 74) the two sides containing the right angle are 3 m and 8 m in length. What is the length of the hypotenuse?

$$a^2 = b^2 + c^2$$
$$= 3^2 + 8^2\ \text{m}^2$$
$$= 9 + 64\ \text{m}^2 = 73\ \text{m}^2$$
$$\therefore a = \sqrt{73}\ \text{m} = \underline{8.54\ \text{m}}$$

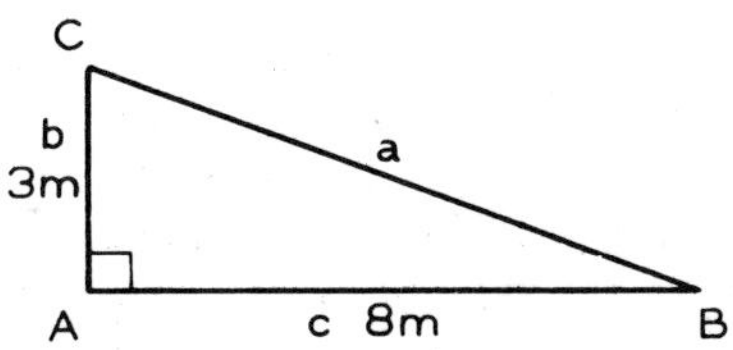

Figure 74

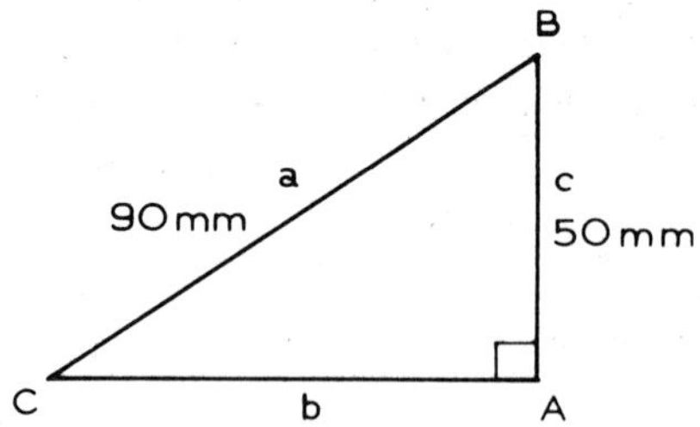

Figure 75

Example Find the side b of the right-angled triangle in Figure 75.

$$a^2 = b^2 + c^2$$

$$\therefore b^2 = a^2 - c^2$$

$$= 90^2 - 50^2 \text{ mm}^2$$

$$= 8\,100 - 2\,500 \text{ mm}^2 = 5\,600 \text{ mm}^2$$

$$\therefore b = \sqrt{5\,600} \text{ mm} = \underline{74.83 \text{ mm}}$$

16.5 Special cases

Right-angled triangles whose three sides are integers are both interesting and useful. Figure 76 shows a right-angled triangle in which the two sides forming the arms of the right angle are 3 units and 4 units (these may be any units).

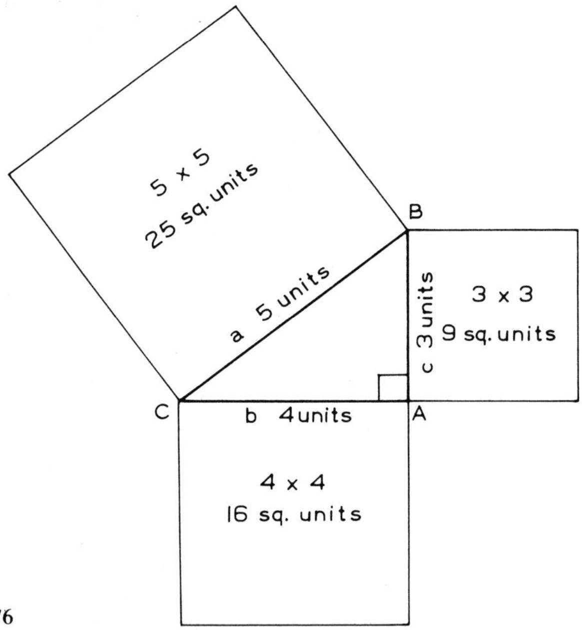

Figure 76

The hypotenuse a is therefore given by:

$$a^2 = b^2 + c^2$$

$$= 4^2 + 3^2$$

$$= 16 + 9 = 25$$

$$\therefore a = \sqrt{25} = 5$$

Such triangles are called 3 : 4 : 5 triangles, and are very useful for the purpose of constructing right angles. Any proportional values to 3 : 4 : 5 may be used, such as 6 : 8 : 10 or 15: 20 : 25 and so on.

A well used site application is to place together the zero and 24 m points of a linen measuring tape, and to pull the resulting loop into a triangle with the 6 m and 16 m marks as the other two vertices. Its sides will be 6 m, 8 m and 10 m, and the angle at the 0, 24 m junction will be a right angle. On a smaller scale a triangle of timber battens may be used. Both are illustrated in Figure 77.

Another convenient right-angled triangle is the 5 : 12 : 13 triangle. Check for yourself that $5^2 + 12^2 = 13^2$.

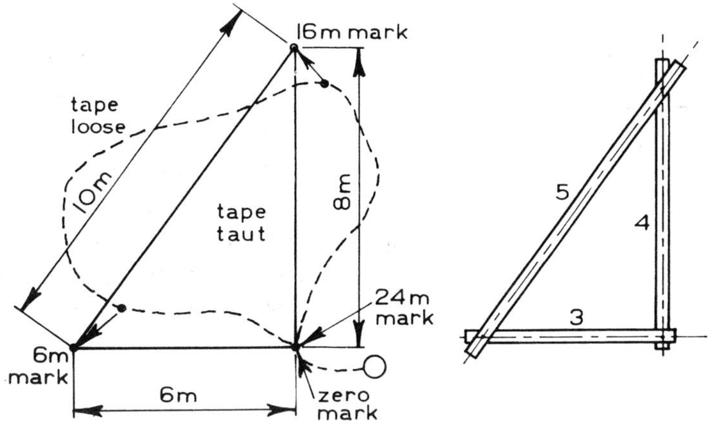

Figure 77

16.6 Similar triangles and congruent triangles

If two triangles have their corresponding angles equal they are said to be *similar triangles.*

In Figure 78, ABC and ADE have a common angle at A.

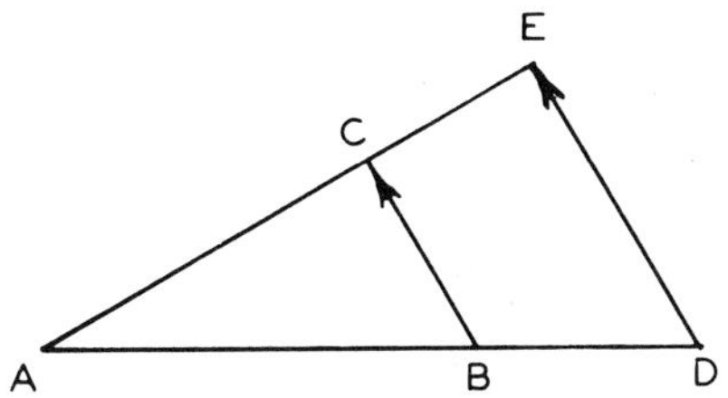

Figure 78

$\angle BCA = \angle DEA$ (corresponding angles)

$\angle ABC = \angle ADE$ (corresponding angles)

$\therefore$ ABC and ADE are similar triangles

The corresponding sides of similar triangles are proportional; i.e. in Figure 78,

$$\frac{AC}{AE} = \frac{AB}{AD} = \frac{BC}{DE}$$

If similar triangles also have equal dimensions, i.e. are equal in all respects, they are said to be *congruent.* Conditions of congruency are:
i) two corresponding sides and the included angle equal,
ii) three sides equal,
iii) two angles and one corresponding side equal.
In right-angled triangles:
iv) the hypotenuse and one corresponding side equal.

16.7 Use of similar triangles

Similarity and congruency often provide means of solving problems in geometry.

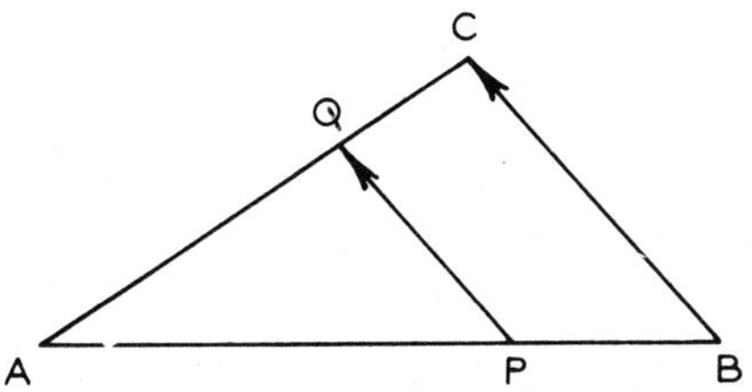

Figure 79

Example ABC (Figure 79) is a triangle, and PQ is a line parallel to BC cutting AB at P and AC at Q. AB = 70 mm, AC = 50 mm and AP = 40 mm. Find the length AQ.

APQ and ABC are similar triangles (see section 16.6).

$$\therefore \frac{AQ}{AC} = \frac{AP}{AB}$$

$$\therefore \ AQ = \frac{AC.AP}{AB}$$

$$= \frac{50 \times 40}{70} \text{ mm} = \underline{28.6 \text{ mm}}$$

Example The distance between two points, A and B, is required but the space between them is inaccessible (Figure 80). How can the distance AB be found?

A line BD is set out at right angles to AB, and a second line DQ at right angles to BD. The mid-length of BD is marked P. The point Q is now selected such that QPA is a straight line. In triangles ABP and DPQ:

$\angle APB = \angle DPQ$ (vertically opposite angles)

$\angle ABP = \angle PDQ$ (right angles)

$BP = PD$

Therefore the triangles are congruent, i.e.

$DQ = AB$

Measure DQ to obtain the distance AB.

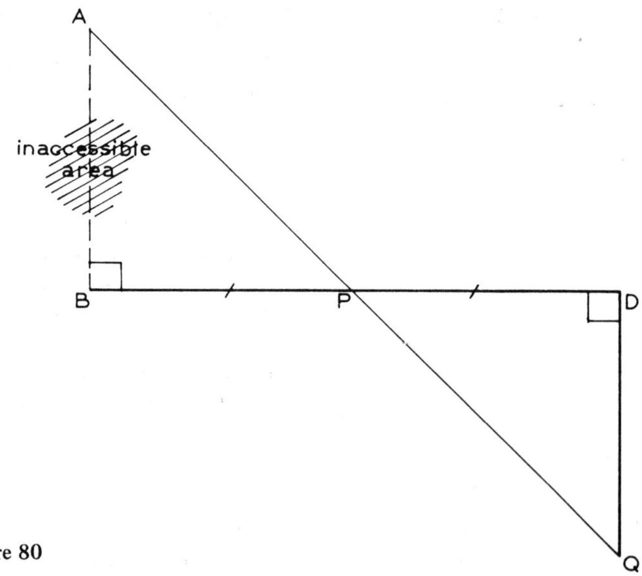

Figure 80

Example Two points A and B are on opposite sides of a building complex, as shown in Figure 81, and the distance between them is required.

Set out two straight lines, one from A and one from B to intersect at C. Mark D at the mid-point of AC, and E at the mid-point of BC.

In triangles ABC and DEC:

$\angle ACB$ is common

$$\frac{DC}{AC} = \frac{1}{2} \qquad \frac{EC}{BC} = \frac{1}{2}$$

Therefore the triangles are similar, i.e.

$$\frac{DE}{AB} = \frac{1}{2}$$

$\therefore$ DE $= \frac{1}{2}$AB

Measure DE and calculate AB.

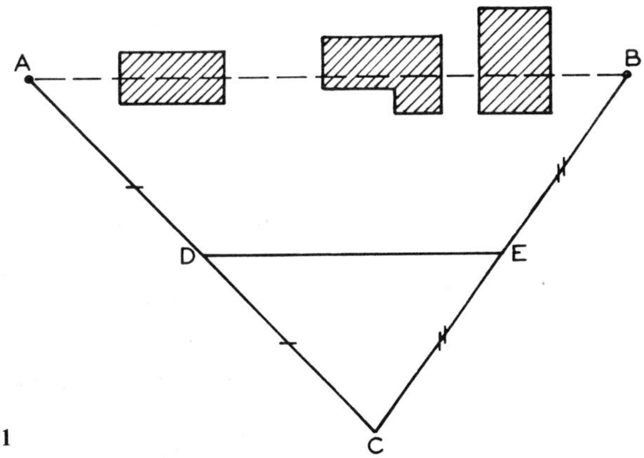

Figure 81

16.8 Construction of triangles

The following examples illustrate the construction of triangles from given information.

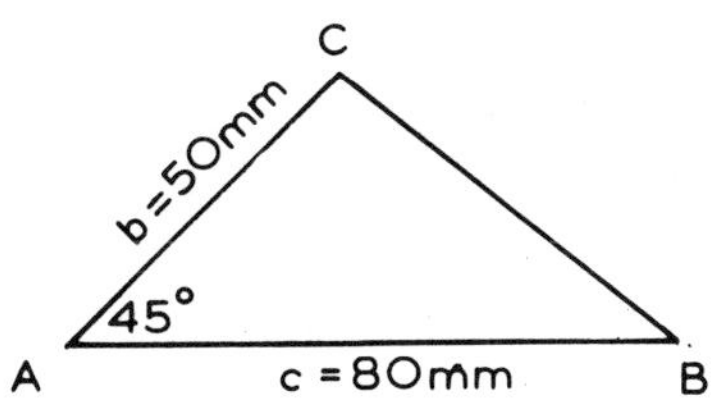

Figure 82

Example Construct the triangle $\angle A = 45°$, $b = 50$ mm, $c = 80$ mm (these are two sides and the included angle).

Draw AB = 80 mm (side c).
Draw AC = 50 mm at 45° to AB (side b).
Join BC.
ABC is the required triangle (Figure 82).

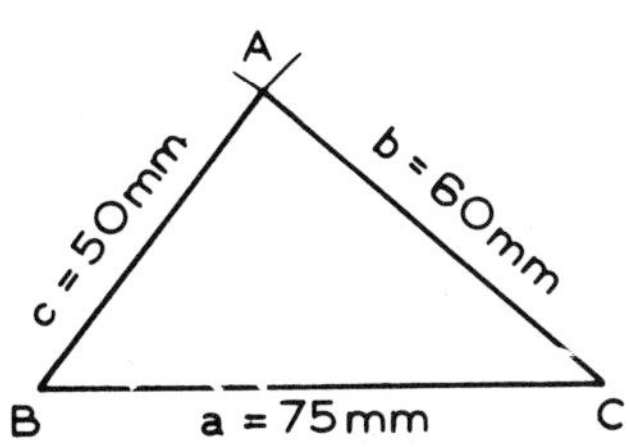

Figure 83

Example Construct the triangle $a = 75$ mm, $b = 60$ mm, $c = 50$ mm (three sides given).

Draw BC = 75 mm (side a).
With centre B, describe an arc of radius 50 mm.
With centre C, describe an arc of radius 60 mm.
Mark the point of intersection of the arcs A.
Join AB and AC.
ABC is the required triangle (Figure 83).

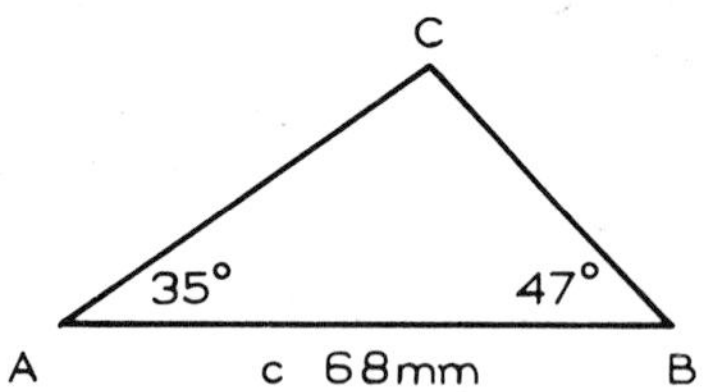

Figure 84

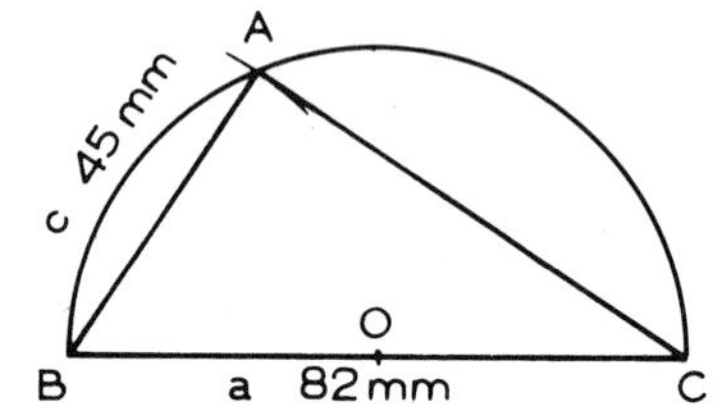

Figure 85

Example Construct the triangle $\angle A = 35°$, $\angle B = 47°$, $c = 68$ mm.

Draw AB = 68 mm (side c).
Draw AC at 35° to AB and BC at 47° to AC (using a protractor).
ABC is the required triangle (Figure 84).

Example Construct the right-angled triangle having the hypotenuse of length 82 mm and another side of length 45 mm.

Draw BC = 82 mm (side a).
Bisect BC at O and construct a semicircle, centre O, on BC.
With centre B and radius 45 mm, describe an arc to cut the semicircle at A.
Join AB and AC.
Then $\angle BAC = 90°$ and ABC is the required triangle (Figure 85).

Note. The basis of this construction is that the angle in a semicircle is a right angle; see section 17.3.

Exercise 16b

1. Find the third side of each of the following right-angled triangles, the right angle being contained by the given sides in each case.

a) 50 m, 12 m; b) 25 mm, 60 mm; c) 80 mm, 150 mm;
d) 5.5 m, 6.8 m.

2. What is the length of the diagonal of a rectangle that has sides of length 3 m and 2.25 m?

3. The diagonal of a rectangle is 125 mm and one side is 75 mm. Find its other dimension.

4. A rectangular box is 1 m long. Its breadth and depth are respectively a third and a quarter of its length. What is the longest straight wire that can be contained within the box?

5. A ladder 8 m long just reaches the top of a wall 5.75 m high. How far from the foot of the wall is the foot of the ladder?

6. The longest side of a triangle is 30 m. What are its other two sides if it is similar to a triangle having sides of 2 m, 4 m and 5 m?

7. There are twenty-four equally spaced rungs in a ladder, which leans against a wall with its third rung 375 mm above the ground. What is the height of the top rung?

8. Two men stand side by side, and their shadows are measured as 3.1 m and 2.8 m. If the taller man is 1.8 m in height, what is the height of the other?

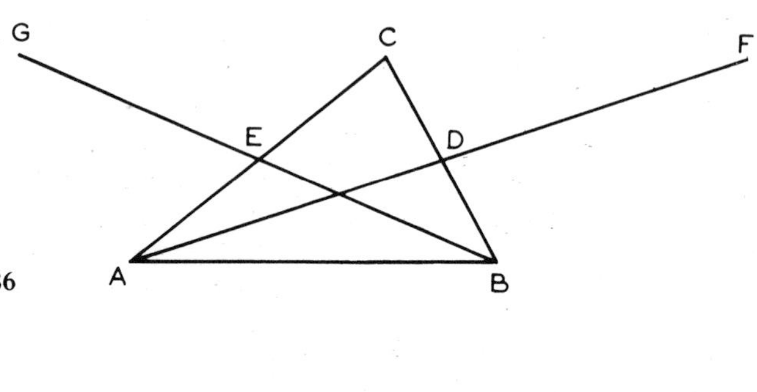

Figure 86

9. Figure 86 shows a triangle ABC in which the vertices A and B are joined to the mid-points of the opposite sides, D and E respectively. AD is produced to F so that DF = AD, and BE is produced to G so that EG = BE. Complete the triangle CGE, and show that it is congruent with ABE.

10. In Figure 86, find a triangle which is congruent with ABD.

11. Construct the following triangles and measure the remaining sides (nearest mm) and angles.

a) a = 65 mm, b = 70 mm, c = 85 mm
b) a = 50 mm, $\angle B = 45°$, $\angle C = 60°$
c) a = 57 mm, b = 68 mm, $\angle C = 100°$
d) $\angle A = 90°$, b = 37 mm, c = 54 mm
e) $\angle A = 90°$, a = 70 mm, b = 37 mm

17 CIRCLES

A circle is a plane figure contained by a line traced out by a point moving at a constant distance from a fixed point.

17.1 Parts of a circle

The fixed point is the *centre*, and the line traced out by the moving point is the *circumference* of the circle. The constant distance between the two points is the *radius* of the circle, and any straight line through the centre joining two points on the circumference is a *diameter.*

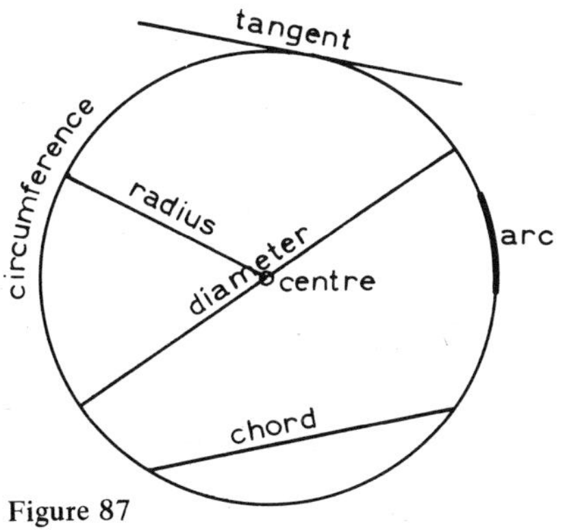

Figure 87

Diameter = 2 × radius

Any straight line joining two points on the circumference but not passing through the centre is a *chord.* Any part of the circumference is an *arc.* A straight line outside the circle which touches the circumference at one point only is a *tangent* to the circle. All the above parts of the circle are illustrated in Figure 87.

A *sector* is a portion of a circle bounded by an arc and two radii. A *segment* is bounded by an arc and a chord. Both are shown in Figure 88. Note that a chord divides a circle into two parts, a *major segment* and a *minor segment*. Similarly, a pair of radii divide the circle into a *major sector* and a *minor sector.*

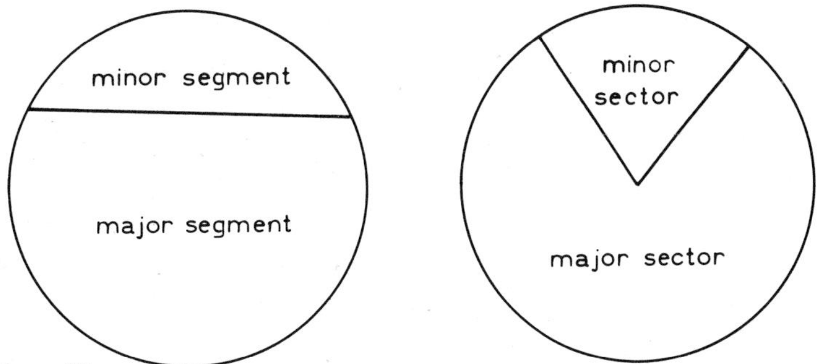

Figure 88

17.2 The constant π

Careful measurement of the diameters and circumferences of a number of different sized circles will show that the ratio circumference/diameter is a constant. The Greek letter π is used to denote this constant, the value of which is 3.142 to four significant figures. It is in fact a non-terminating, non-recurring decimal. When it is convenient to use vulgar fractions, π may be taken as $3\frac{1}{7}$.

This relationship between circumference and diameter is most commonly expressed in the formula:

$C = \pi d$

It may also be expressed in the form:

$C = 2\pi r$

since $d = 2r$.

Example The diameter of a circle is 325 mm. What is its circumference?

$C = \pi d$

$= \pi \times 325\text{ mm}$

$= \underline{1\,021\text{ mm}}$ (to 4 significant figures)

Example A circular loop of wire is re-shaped to form a perfect square. What is the size of the square?

Assuming no stretching occurs:

Perimeter of square = circumference of circle

i.e. $4s = 2\pi r$

where s is the length of the side of the square.

$\therefore s = \frac{1}{4} \times 2\pi r$

$= \frac{1}{4} \times 2\pi \times 300\text{mm}$

$= 150\pi\text{ mm}$

$= \underline{471.2\text{ mm}}$

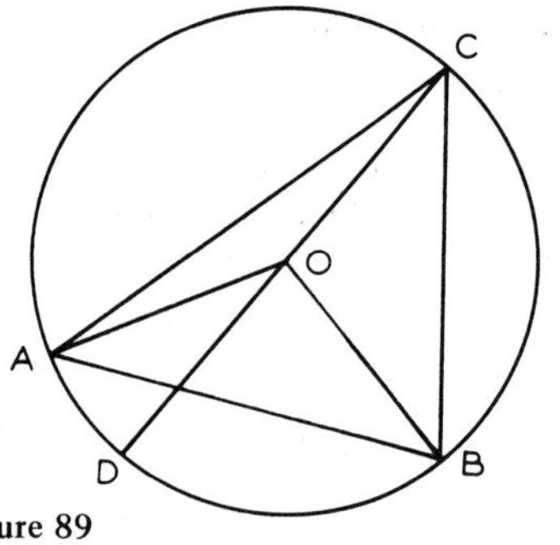

Figure 89

Example The girth of a circular concrete pipe is 4.5 m and its wall thickness is 30 mm. What is its internal diameter?

First find the external diameter.

$C = \pi d$

$\therefore d = C/\pi$

$= 4\,500/\pi\text{ mm}$

$= 1\,432\text{ mm}$

$\therefore$ Internal diameter $= 1\,432 - 2(30) = \underline{1\,372\text{ mm}}$

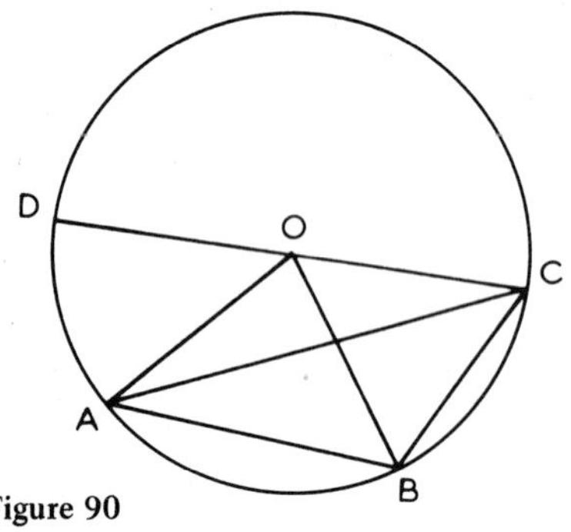

Figure 90

17.3 Angles in circles

In Figures 89 and 90, ∠AOB and ∠ACB are the angles at the centre and circumference, respectively, standing on arc AB. It may be shown that:

∠AOB = 2 × ∠ACB

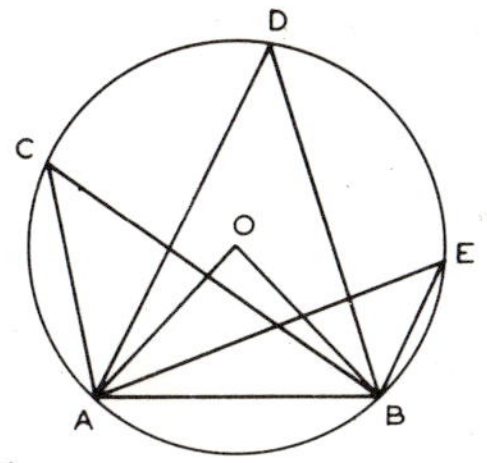

Figure 91

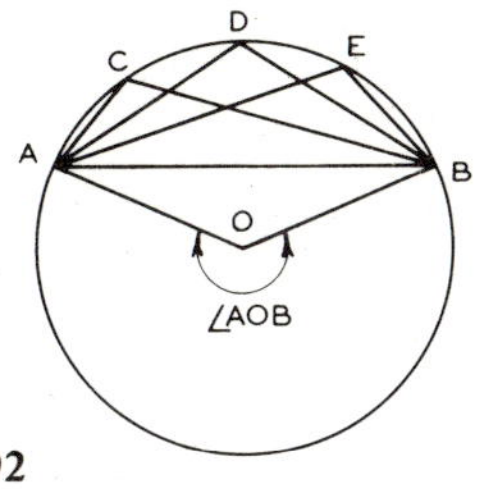

Figure 92

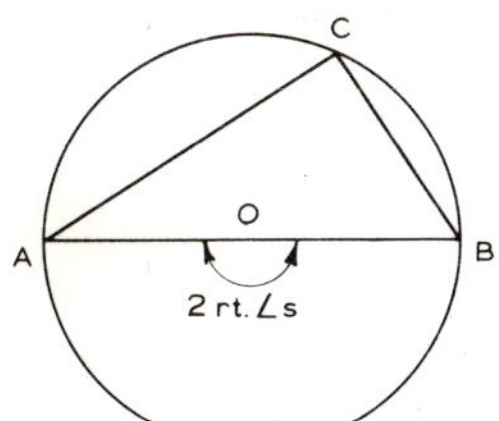

Figure 93

Proof Join CO and produce to D. In triangle OCA:

$$OA = OC \quad \text{(radii)}$$

$$\therefore \angle OCA = \angle OAC$$

$$\therefore \angle OCA + \angle OAC = 2\angle OCA$$

$$\text{Exterior } \angle AOD = \angle OCA + \angle OAC \quad \text{(see section 16.1)}$$

$$= 2\angle OCA$$

$$\text{Similarly: } \angle BOD = 2\angle OCB$$

Therefore, by addition in Figure 89 and subtraction in Figure 90,

$$\underline{\angle AOB = 2\angle ACB}$$

In words, *the angle at the centre of a circle is twice the angle at the circumference standing on the same arc.*

It follows that *angles in the same segment of a circle are equal.* Referring to Figures 91 and 92, ∠AOB is the angle at the centre.

$$\therefore \angle ACB = \tfrac{1}{2}\angle AOB$$

$$\angle ADB = \tfrac{1}{2}\angle AOB$$

$$\text{and } \angle AEB = \tfrac{1}{2}\angle AOB$$

$$\therefore \angle ACB = \angle ADB = \angle AEB$$

and so on for any angle in the same segment.

Note. In Figure 92 the arc AB is a major arc, and the angle it subtends at the centre is the *reflex* angle AOB.

The angle in a semicircle is a right angle (see Figure 93). *Angles in minor segments are obtuse; angles in major segments are acute.*

17.4 Angle between tangent and radius

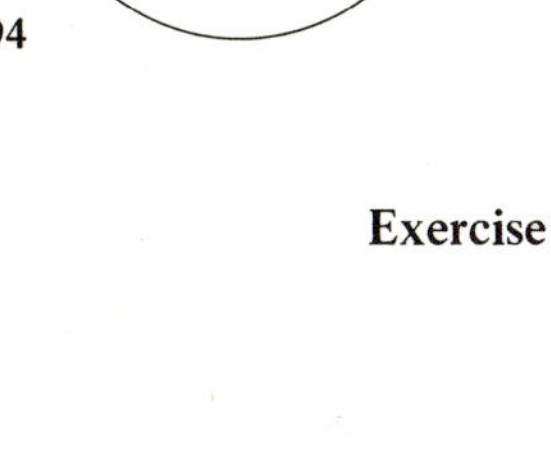

Figure 94

In Figure 94, P is the point of intersection of the tangent AB and the radius OP of circle, centre O. Q and R are points on AB equidistant from P.

$$\therefore \angle OQP = \angle ORP$$

If Q and R move nearer to P, remaining equidistant from it, angles OQP and ORP increase until, when P, Q and R coincide:

$$\underline{\angle OPA = \angle OPB = 90^\circ}$$

i.e. *the angle between tangent and radius is a right angle.*

Exercise 17a

1. A circle drawn with compasses has a circumference of 1 571 mm. To what radius are the compasses set?

2. A wheel of diameter 550 mm spins at 150 revolutions per minute. How far does a point on the circumference travel in one second?

3. What is the diameter of a wheel that makes 100 turns while travelling 110 m?

4. How many revolutions will be made by a millstone of diameter 1.25 m in travelling through one circuit of radius 1.5 m?

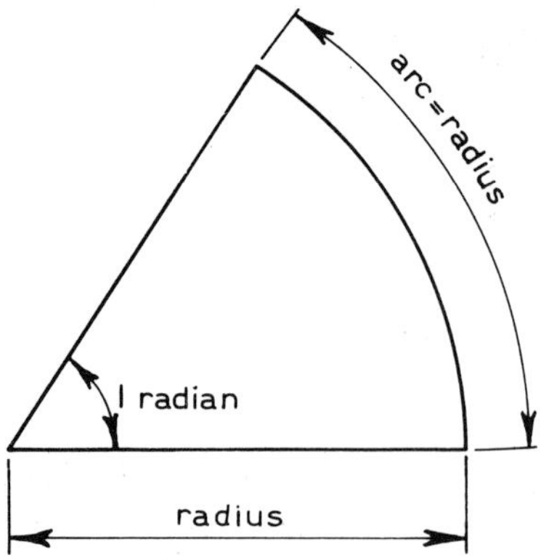

Figure 95

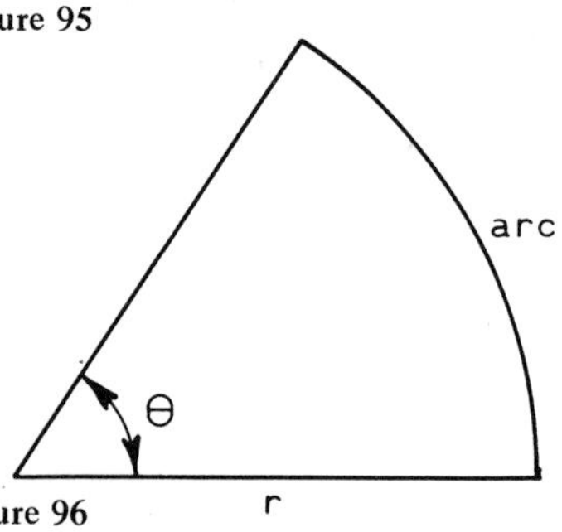

Figure 96

5. In a circle, centre O, a chord AB is drawn equal in length to the radius. What are the angles of the triangle OAB?

6. APB and CPD are chords of a circle intersecting at P. ∠PAC = 30° and ∠PCA = 45°. Find the angles PBD and PDB.

7. AB and CD are chords of a circle which, when produced, meet at a point P outside the circle. ∠ABC = 25°, ∠CPB = 40°. Find the angles ADC and PAD.

8. A, B and C are points on the circumference of a circle, centre O, such that ∠BAC = 75°. Find the angles BOC and OBC.

9. AB and CD are chords of a circle of which AC is a diameter. ∠BAC = 40° and ∠ACD = 55°. Find angles BDC and CBD.

10. A tangent is drawn to a circle of radius 150 mm from a point 200 mm distant from the circle's centre. What is the length of the tangent?

11. Two concentric circles have diameters of 2.6 m and 1 m. AB is a chord of the larger circle and a tangent to the smaller. Find the length of AB.

12. Two circles have their centres 50 mm apart. One has a radius of 15 mm, the other of 30 mm. Draw the circles and their two common tangents. Find the lengths of the tangents.

17.5 Radian measure

A *radian* is the angle subtended at the centre of a circle by an arc that is equal in length to the radius (Figure 95). Angles may be measured in radians as an alternative to measurement in degrees.

In Figure 96, the value of θ in radians is equal to the number of times the radius is contained in the arc.

17.6 Conversion between degrees and radians

Since the radius is contained 2π times in the circumference it follows that one revolution contains 2π radians (Figure 97).

i.e. $360^\circ = 2\pi$ radians

$\therefore\ 1^\circ = \dfrac{2\pi}{360}$ radians

$1 \text{ radian} = \dfrac{360^\circ}{2\pi}$

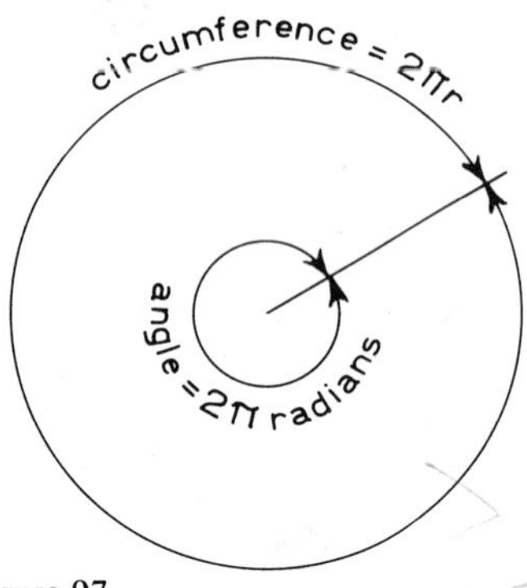

Figure 97

Example Express the angle 35° 30′ in radians.

$$35^\circ\,30' = 35\tfrac{1}{2}^\circ = 35\tfrac{1}{2} \times \frac{2\pi}{360} \text{ radians}$$

$$= \frac{71}{2} \times \frac{2\pi}{360} \text{ radians}$$

$$= \underline{0.6196 \text{ radian}}$$

Example Convert 3.4 radians to degrees and minutes.

$$3.4 \text{ radians} = 3.4 \times \frac{360}{2\pi}^{\circ}$$

$$= 194.8^{\circ} = \underline{194^{\circ}\,48}$$

Many mathematical tables contain conversions between degrees and radians to save laborious calculations.

17.7 Common angles in radians

It is not unusual to express angular displacement in radians in terms of π. From the fact that 2π radians = 360° it is a simple matter to construct the following table of the common angles in similar terms.

Degrees	30	45	60	90	180
Radians	$\pi/6$	$\pi/4$	$\pi/3$	$\pi/2$	π

17.8 Arcs and angles

By the definition of a radian:

arc = radius × angle in radians

i.e. arc = $r\theta$

where θ is in radians. See Figure 98.

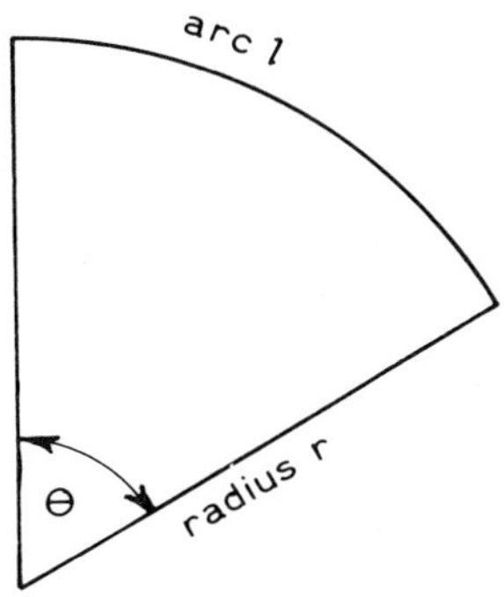

Figure 98

Example What is the length of an arc that subtends an angle of 40° in a circle of 1.25 m radius?

First express the angle in radians.

$$40^{\circ} = 40 \times \frac{2\pi}{360} = \frac{2\pi}{9} \text{ radians}$$

$$\text{arc} = r\theta = 1.25 \times \frac{2\pi}{9} = \underline{0.873 \text{ m}}$$

Example What is the angle subtended by a circular arc of length 3.6 m on a radius of 6.5 m?

arc = $r\theta$

$$\therefore \theta = \frac{\text{arc}}{r} = \frac{3.6}{6.5} = \underline{0.554 \text{ radian}}$$

Example A pendulum swinging through an arc of $\pi/9$ radians travels a distance of 160 mm in one complete swing. What is the length of the pendulum?
Note. One complete swing is the motion from one position to the other and back, i.e. two arcs completed. ∴ arc = 80 mm.

arc = $r\theta$

$$\therefore r = \frac{\text{arc}}{\theta} = \frac{80}{\pi/9} = \frac{720}{\pi} = \underline{229.2 \text{ mm}}$$

Exercise 17b

1. Find the angle in degrees and minutes subtended at the centre of a circle by an arc equal in length to the radius.

2. Convert the following angles to radians, correct to four significant figures:

a) $37^\circ 30'$ b) 175° c) $210^\circ 36'$

3. Convert the following angles to degrees and minutes correct to the nearest minute:

a) 0.875 radians b) 3.63 radians c) 5.72 radians

4. Express the following angles in radians in terms of π:

a) 10° b) 15° c) 36° d) 100° e) 225°

5. Convert the following to degrees:

a) $3\pi/4$ radians b) $\pi/15$ radians c) $2\pi/5$ radians

6. A circular arc subtends an angle of 48° at the centre of a circle of 2.75 m radius. What is the length of the arc?

7. In a circle of diameter 7.6 m, a pair of radii cut off an arc of length 4.75 m. What is the figure thus formed? What is the angle between the two radii: a) in radians, b) in degrees and minutes?

8. A hand rail 4 m long is to be bent to form a circular curve on a radius of 5.5 m. Through what angle must it be bent?

9. The swing of a pendulum allows the bob to travel a distance of 225 mm in one complete movement. If the pendulum is 1.15 m in length, what is the angle of swing: a) in radians, b) in degrees and minutes?

10. Through what angle does a vehicle turn when negotiating a circular curve of length 124 m at a radius of 55 m? Give the answer in radians and in degrees and minutes.

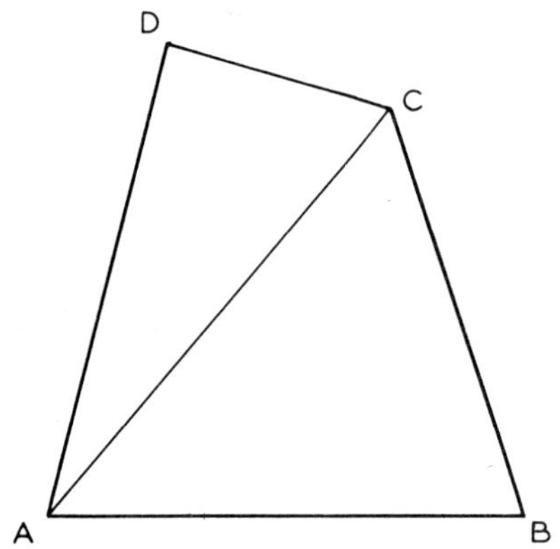

Figure 99

18 AREAS AND VOLUMES

18.1 Triangulation

Rectilinear figures are those bounded by straight lines, and any such figure may be divided into triangles. This *triangulation* is an important aspect of land surveying techniques and structural frame design.

A four-sided rectilinear figure is a *quadrilateral.* Figure 99 shows a quadrilateral divided into two triangles by one of its diagonals. Since the internal angles of each triangle have the sum 180°, the sum of the angles of the quadrilateral is 360°.

18.2 Quadrilaterals

Rectangle (Figure 100)

Each internal angle is a right angle.
Opposite sides are parallel and equal.
Adjacent sides are unequal.
Diagonals are equal and bisect one another.

Square (Figure 101)

Each internal angle is a right angle.
All sides are equal.
Opposite sides are parallel.
Diagonals are equal and bisect one another at right angles.

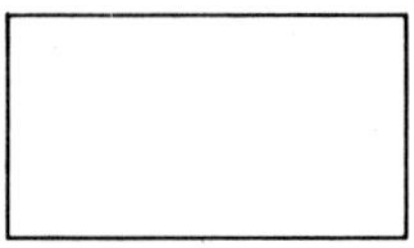

Figure 100

Figure 101

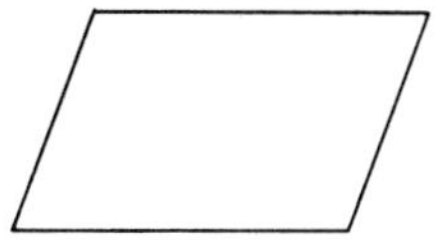

Figure 102

Parallelogram (Figure 102)

Opposite sides are parallel and equal.
Adjacent sides are unequal.
Opposite angles are equal.
Adjacent angles are supplementary.
Diagonals are unequal and bisect one another.

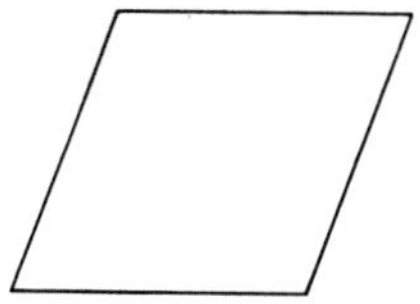

Figure 103

Rhombus (Figure 103)

A parallelogram with all sides equal.
Diagonals bisect one another at right angles.

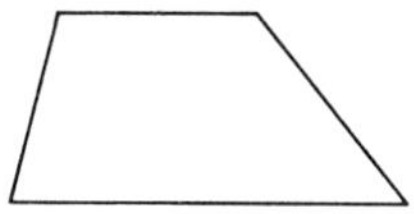

Figure 104

Trapezium (Figure 104)

Any quadrilateral with one pair of parallel sides.

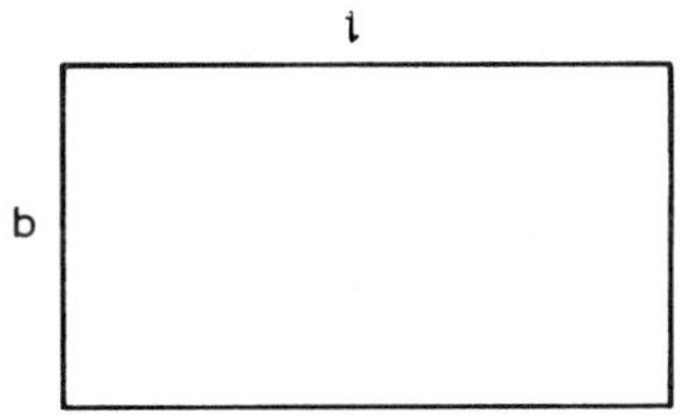

Figure 105

18.3 Areas of plane figures

The unit of area, or superficial measure, is the square with side of unit length.

For a square with side s units, area is given by:

$A = s^2$

and perimeter is given by:

$P = 4s$

Rectangle (Figure 105)

Area $A = lb$

Perimeter $P = 2(l + b)$

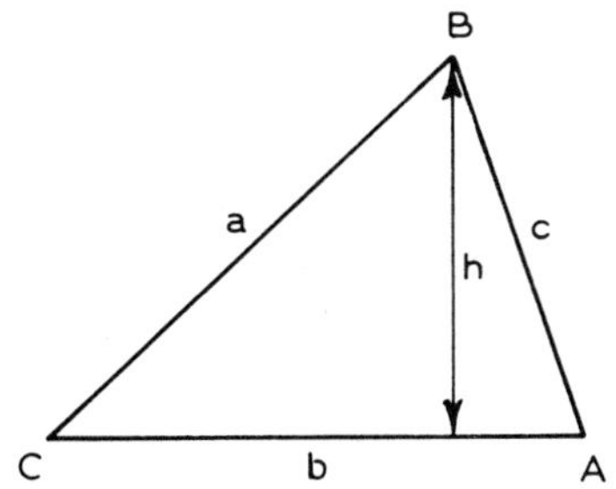

Figure 106

Triangle A triangle is half of the rectangle on the same base and having the same perpendicular height (Figure 106).

Area $A = \frac{1}{2}bh$

Perimeter $P = a + b + c$

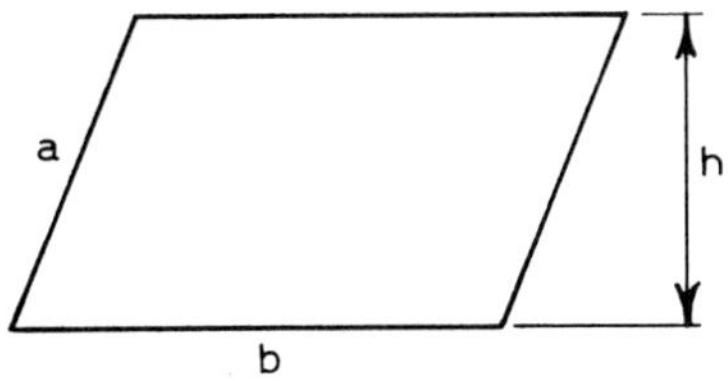

Figure 107

Parallelogram A parallelogram is equal in area to a rectangle on the same base and having the same perpendicular height (Figure 107).

Area $A = bh$

Perimeter $P = 2(a + b)$

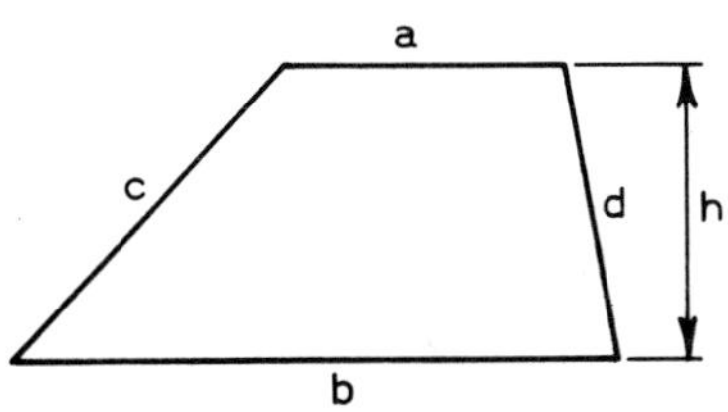

Figure 108

Trapezium The trapezium may be related to a rectangle whose length is the mean of the parallel sides of the trapezium (Figure 108).

Area $A = \frac{1}{2}h(a + b)$

Perimeter $P = a + b + c + d$

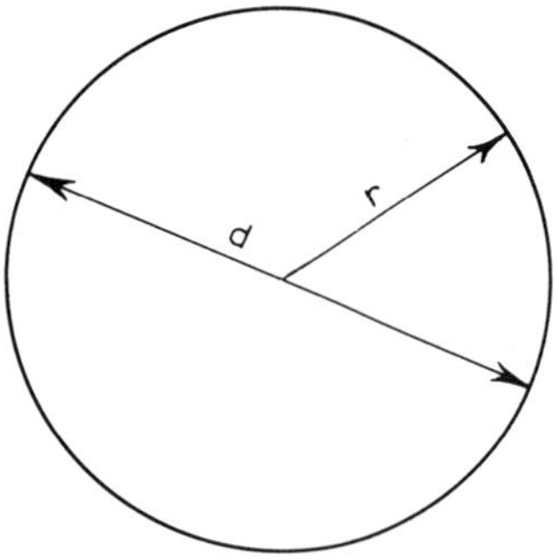

Figure 109

Circle (Figure 109)

$$\text{Area } A = \pi r^2 \quad (\text{or } \tfrac{1}{4}\pi d^2)$$

$$\text{Perimeter } C = 2\pi r \quad (\text{or } \pi d)$$

Example A square field has an area of 9 604 m². What is the length of a path that runs diagonally across it (Figure 110)?

$$\text{Length of side AB} = \sqrt{9\,604}$$
$$= 98 \text{ m}$$
$$\text{Length of diagonal AC} = \sqrt{(98^2 + 98^2)}$$
$$= \sqrt{(2 \times 98^2)}$$
$$= 98\sqrt{2}$$
$$= \underline{138.6 \text{ m}}$$

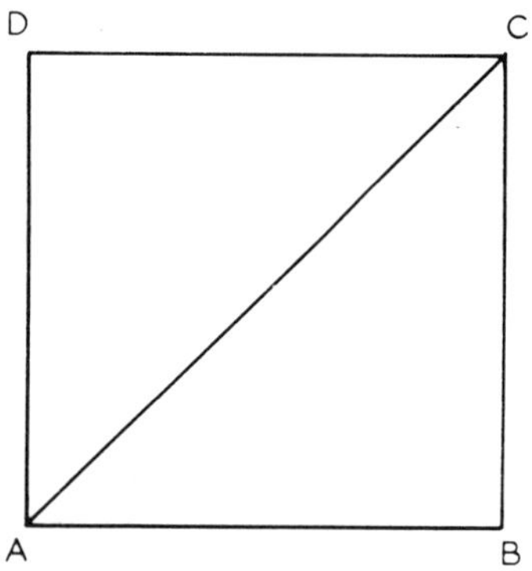

Figure 110

Example Find the cross-sectional area of the section shown in Figure 111.

Inspection will show that the required section is a rectangle 70 mm × 60 mm, minus a square of side 50 mm.

$$\therefore \text{Area } A = (70 \times 60) - 50^2 \text{ mm}^2$$
$$= 4\,200 - 2\,500 \text{ mm}^2$$
$$= \underline{1\,700 \text{ mm}^2}$$

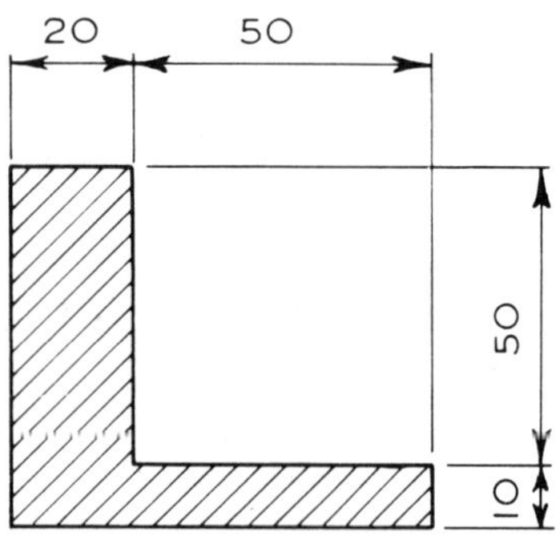

Figure 111

Example A parallelogram has one pair of opposite sides 154 mm long, and the perpendicular distance between them is 83 mm (Figure 112). What is its area?

$$\text{Area } A = bh$$
$$= 154 \times 83 \text{ mm}^2$$
$$= \underline{12\,782 \text{ mm}^2}$$

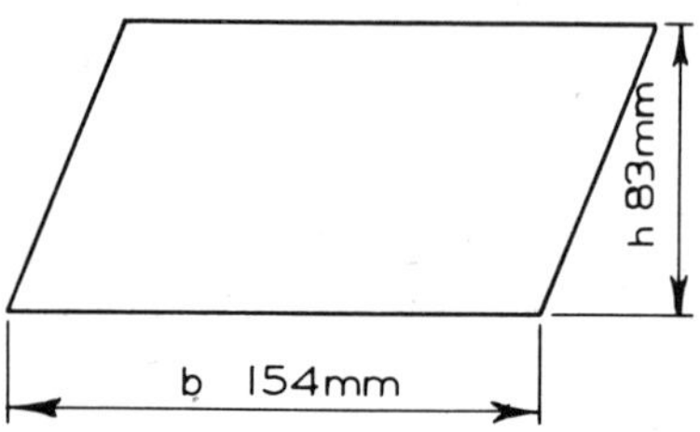

Figure 112

Example Figure 113 shows the cross-section of a trapezoidal retaining wall. Find the cross-sectional area in square metres.

$$\text{Area } A = \tfrac{1}{2}h(a + b)$$
$$= \tfrac{1}{2} \times 3.5(1 + 2.4) \text{ m}^2$$
$$= \tfrac{1}{2} \times 3.5 \times 3.4 \text{ m}^2$$
$$= \underline{5.95 \text{ m}^2}$$

Example A circle is drawn through the four corners of a square that has a side of 6 m (Figure 114). Find the area between the square and the circle.

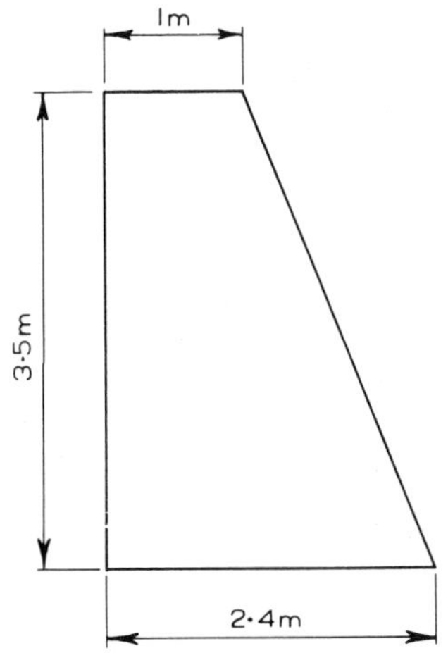

Figure 113

$$\text{Required area} = \text{circle} - \text{square}$$
$$= \pi r^2 - s^2$$
$$= \tfrac{1}{2}\pi s^2 - s^2 \quad \text{(see Note)}$$
$$= s^2(\tfrac{1}{2}\pi - 1)$$
$$= 6^2(\tfrac{1}{2} \times 3.142 - 1)\,\text{m}^2$$
$$= 36 \times 0.571\,\text{m}^2$$
$$= \underline{20.56\,\text{m}^2}$$

Note $r^2 = (\tfrac{1}{2}s)^2 + (\tfrac{1}{2}s)^2$
$$= \tfrac{1}{4}s^2 + \tfrac{1}{4}s^2$$
$$= 2(\tfrac{1}{4}s^2)$$
$$= \tfrac{1}{2}s^2$$

Exercise 18a Construct and name the following quadrilaterals. In each case the letters A, B, C, D read anticlockwise round the figure.

1. AB = CD = 52 mm, BC = AD = 32 mm, BD = 72 mm

2. AB = 76 mm, AD = CD = 38 mm, ∠BAD = 90°, ∠ABC = 45°

3. AB = CD = BC = AD = 55 mm, ∠BAD = 55°

4. AB = BC = CD = AD = BD = 60 mm

5. AB = CD = 52 mm, AD = BC = 39 mm, BD = 65 mm

6. One pair of opposite sides of a parallelogram have lengths of 2.5 m and the perpendicular distance between them is 1.75 m. What is the length of side of a square having an area equal to that of the parallelogram?

7. The diagonals of a rhombus are 60 mm and 80 mm in length. What length are its sides?

8. The internal angles of a rhombus, the sides of which are 1.5 m, are 60° and 120°. Find the lengths of its diagonals and its area.

9. A circular lawn 14 m in diameter is laid in a square courtyard having a dimension of 21 m, and the surrounding area is paved. Find the area of paving.

10. The parallel sides of a trapezium are 27 m and 35 m and the perpendicular distance between them is 32 m. Find its area in square metres.

11. The gable end of a brick building is 4.8 m wide and has a maximum height of 7.8 m. The roof is symmetrically pitched at 45°. The wall contains four window openings each 1350 mm wide and 900 mm high. What is the area of brickwork?

12. Find the areas of the following circles: a) radius 1.25 m; b) diameter 7.84 m; c) circumference 15 m.

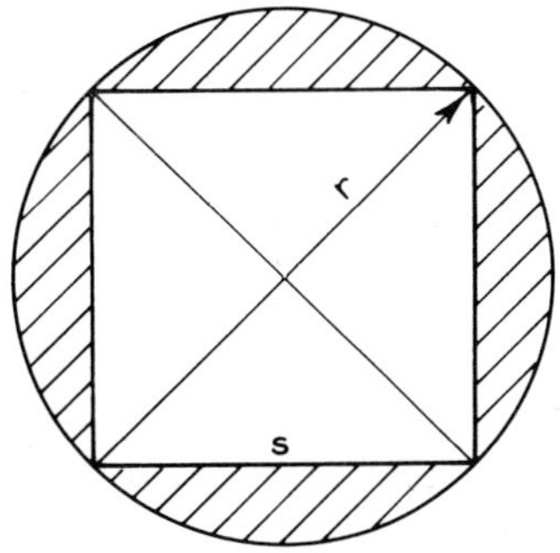

Figure 114

18.4 Volumes of solids

Prisms A prism is a solid having a constant cross-sectional shape and size throughout its length. The name of the prism is taken from the shape of the cross-section; examples are shown in Figure 115. In all cases:

Volume of prism = area of cross-section × length

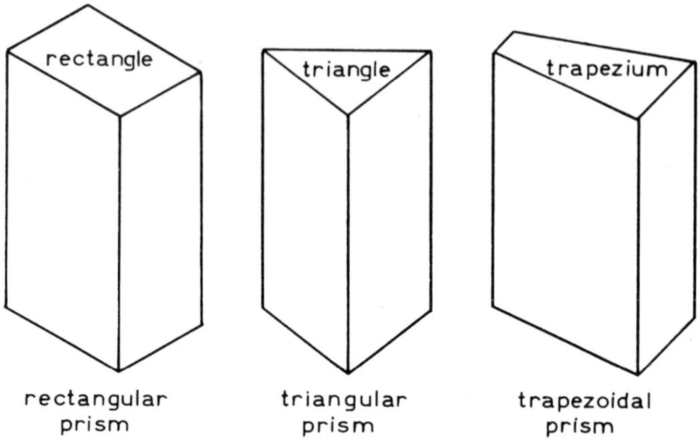

Figure 115

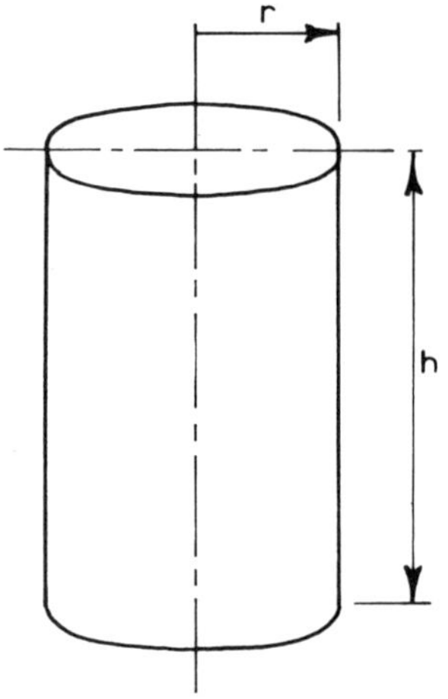

Figure 116

Cylinder A cylinder is a circular prism (Figure 116). Applying the above rule for the volume, we have:

Volume = area of circle × length

or $V = \pi r^2 h$

Pyramids A pyramid may be defined as the solid generated by a straight line, one end of which is fixed while the other end traces out the perimeter of a plane figure. Pyramids are defined by the shape of the base; see Figure 117 for examples. In each case the volume of the pyramid is one third of the volume of the containing prism.

i.e. Volume = $\frac{1}{3}$ base area × height

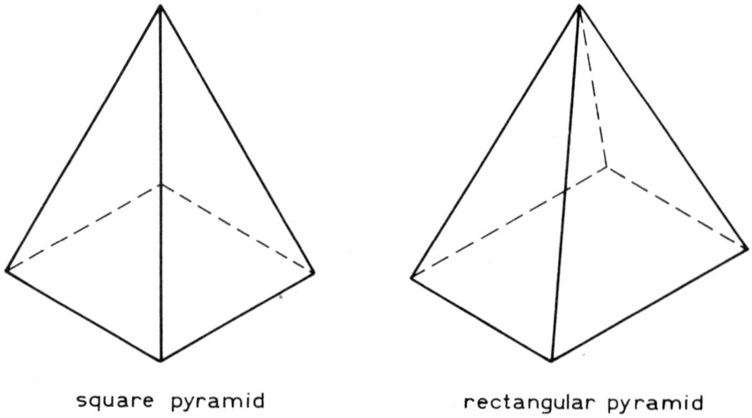

Figure 117

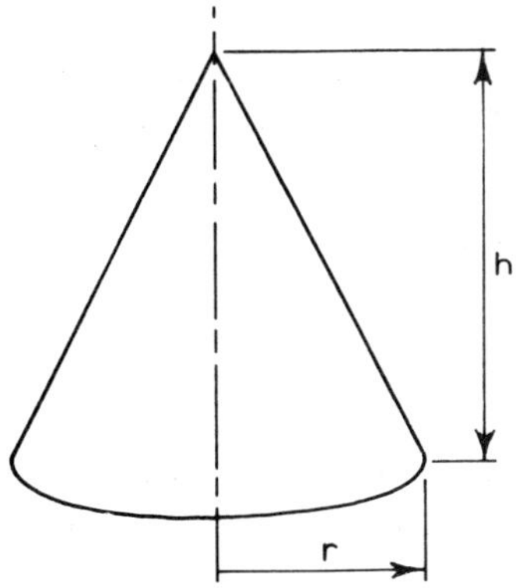

Figure 118

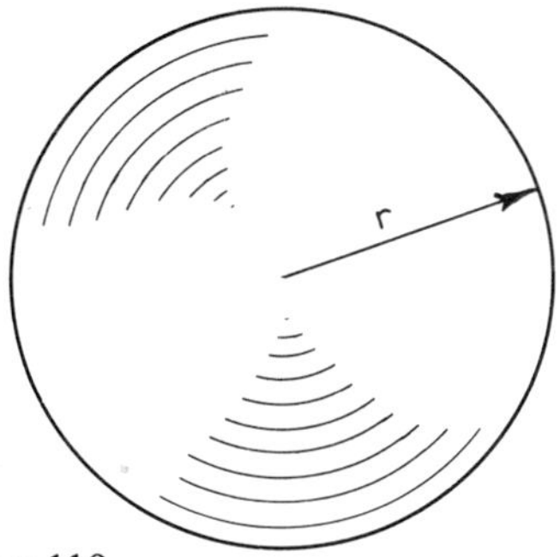

Figure 119

Cone A cone is a circular pyramid (Figure 118). Applying the above rule for volume, we have:

Volume = $\frac{1}{3}$ area of circle × height

or $V = \frac{1}{3}\pi r^2 h$

Sphere A sphere may be defined as the solid generated by a semicircle revolving about its diameter as an axis (Figure 119). The volume of a sphere is given by the formula:

$V = \frac{4}{3}\pi r^3$

Example The cross-section of a triangular prism of hardwood is a 45° isosceles triangle, the equal sides of which are 50 mm long; the length of the prism is 150 mm (Figure 120). What is its volume?

Volume = area of cross-section × length

$= \frac{1}{2} \times 50 \times 50 \times 150\ \text{mm}^3$

$= 187\,500\ \text{mm}^3$ (or $187.5\ \text{cm}^3$)

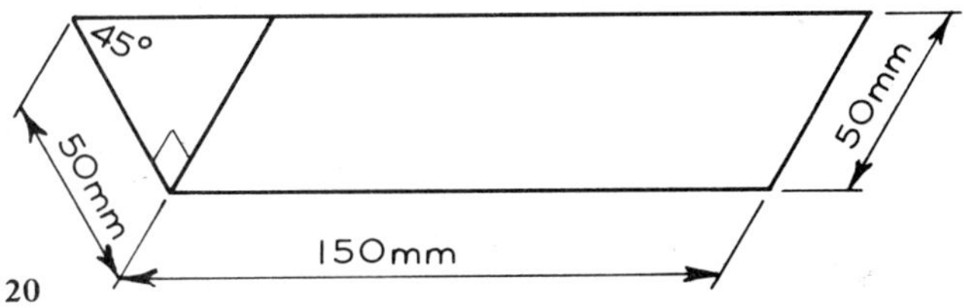

Figure 120

Example A brass cylinder has a diameter of 35 mm and a length of 55 mm. Find its volume in cm^3. If brass has a density of $8.6\ \text{g/cm}^3$, what is the mass of the cylinder?

$V = \pi r^2 h$

$= \pi \times 17.5^2 \times 55\ \text{mm}^3$

or $\pi \times 1.75^2 \times 5.5\ \text{cm}^3 = \underline{52.92\ \text{cm}^3}$

Mass = volume × density

$= 52.92 \times 8.6\ \text{g} = \underline{455.1\ \text{g}}$

Example A concrete pyramid has a square base of side 450 mm and a height at its centre of 300 mm. How much concrete (in cubic metres) is needed to cast a hundred such pyramids?

Volume of each pyramid = $\frac{1}{3}$ base area × height

$= \frac{1}{3} \times 450^2 \times 300\ \text{mm}^3$

$\therefore$ Volume of concrete $= \dfrac{100 \times \frac{1}{3} \times 450^2 \times 300}{10^9}\ \text{m}^3$

$= \underline{2.025\ \text{m}^3}$ (dividing by 10^9 to convert from mm^3 to m^3)

Example A load of sand is tipped into a conical heap with a diameter of 3.5 m and a height of 1.5 m. What volume of sand is contained?

$$V = \tfrac{1}{3}\pi r^2 h$$
$$= \tfrac{1}{3} \times \pi \times 1.75^2 \times 1.5\,\text{m}^3 = \underline{4.81\,\text{m}^3}$$

Example Find the volume of metal in a hollow cast-iron sphere having an external diameter of 70 mm and a wall thickness of 7.5 mm.

This is simply a case of subtracting the volumes, the internal void from the overall volume of the sphere. Denoting the outer radius by R and the inner radius by r:

$$V = \tfrac{4}{3}\pi R^3 - \tfrac{4}{3}\pi r^3$$
$$\text{or } \tfrac{4}{3}\pi(R^3 - r^3) \text{ after factorising}$$
$$= \tfrac{4}{3}\pi(35^3 - 27.5^3)\,\text{mm}^3$$
$$= \tfrac{4}{3}\pi(42\,875 - 20\,797)\,\text{mm}^3$$
$$= \tfrac{4}{3}\pi \times 22\,078\,\text{mm}^3$$
$$= \underline{92\,480\,\text{mm}^3} \quad (\text{or } 92.4\,\text{cm}^3)$$

18.5 Surface areas of solids

The exposed face of a solid will have an area compounded of the areas of the various surfaces it presents.

Prism In the case of prisms, the surface area comprises the two ends plus a number of rectangular sides. For example, the total surface area of the rectangular prism shown in Figure 121 is given by:

$$S = 2lb + 2bh + 2lh$$
$$= 2(lb + bh + lh)$$

Pyramid In the case of pyramids with rectilinear bases, the surface area will be the sum of the base area and a number of triangles forming the sloping sides of the pyramid.

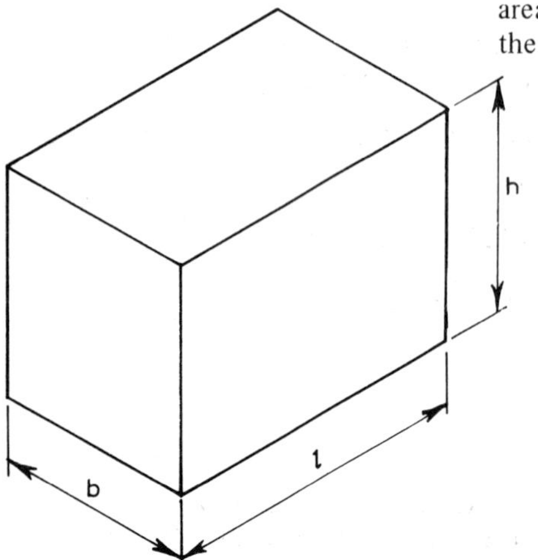

Figure 121

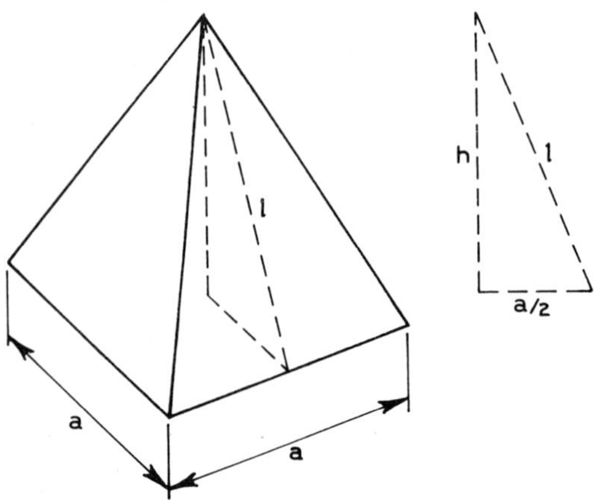

Figure 122

The pyramid in Figure 122 has a square base. Its slant height at the mid points of the sides is denoted by l. Its surface area is given by:

$$S = a^2 + 4(\tfrac{1}{2}al)$$
$$= a^2 + 2al$$
$$= a(a + 2l)$$

Note that the slant height of a pyramid may be found from the base size and the perpendicular height (h), using the theorem of Pythagoras:

$$l^2 = (\tfrac{1}{2}a)^2 + h^2$$

Cylinder The curved surface of a cylinder becomes a rectangle when developed, as illustrated in Figure 123. The area of curved surface is therefore the area of the rectangle.

i.e. Curved surface of cylinder $= 2\pi rh$

To obtain the total surface of the cylinder, it is necessary to add to this the areas of the two end circles. The total is then given by:

$$S = 2\pi rh + 2\pi r^2$$
$$= 2\pi r(h + r)$$

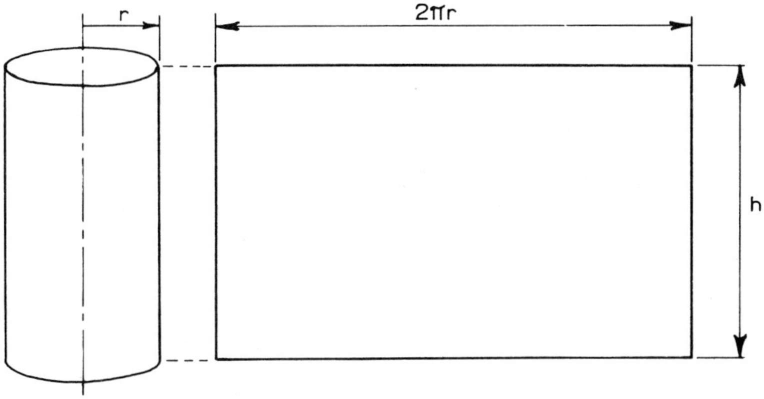

Figure 123

Example Find the area of sheet copper contained in a closed cylinder 525 mm in diameter and 825 mm high.

This means calculating the total surface area of the copper cylinder.

$$S = 2\pi r(h + r)$$
$$= 2 \times \pi \times 262.5 \times (825 + 262.5)\,\text{mm}^2$$
$$= \frac{2 \times \pi \times 262.5 \times 1087.5}{10^6}\,\text{m}^2$$
$$= \underline{1.794\,\text{m}^2}$$

18.6 Composite solids

The following examples illustrate the calculation of the volumes of solids constructed from hemispherical, conical, cylindrical and rectangular parts.

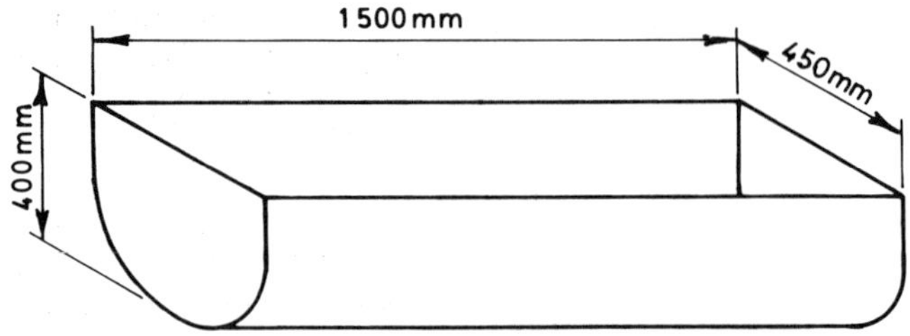

Figure 124

Example A sheet metal trough is rectangular on plan, measuring 450 mm by 1500 mm. Its ends are vertical and formed by a rectangle surmounting a semicircle, the maximum depth being 400 mm. Calculate its capacity in litres and the area of sheet metal it contains.

Figure 124 shows the trough.

End area = semicircle + rectangle

$= \frac{1}{2}\pi r^2 + lb$

$= \frac{1}{2}\pi \times 225^2 + 450 \times 175\ \text{mm}^2$

$= 79\,522 + 78\,750\ \text{mm}^2$

$= 158\,272\ \text{mm}^2$

$= 0.158\,272\ \text{m}^2$

Capacity = end area × length

$= 0.158\,272 \times 1.5\ \text{m}^3$

$= 0.2374\ \text{m}^3$

$= \underline{237.4\ \text{litres}}$

Curved surface (half cylinder) $= \pi rh$

$= \pi \times 225 \times 1\,500\ \text{mm}^2$

$= 1\,060\,288\ \text{mm}^2$

Flat sides area $= 2(175 \times 1\,500)\ \text{mm}^2$

$= 525\,000\ \text{mm}^2$

Total area $= 2(158\,272) + 1\,060\,288 + 525\,000\ \text{mm}^2$

$= 0.316\,544 + 1.060288 + 0.525\ \text{m}^2$

$= \underline{1.9\ \text{m}^2}$

Example A reinforced concrete pile 4.5 m long has a square cross-section of 300 m side, and one end is tapered symmetrically to a point over a length of 400 mm. Find its mass, assuming the density to be 2 400 kg/m³.

Volume of square portion = cross-sectional area × length

$= 0.3 \times 0.3 \times 4.1\,\text{m}^3$

$= 0.369\,\text{m}^3$

Volume of end pyramid = $\frac{1}{3}$base area × length

$= \frac{1}{3} \times 0.3 \times 0.3 \times 0.4\,\text{m}^3$

$= 0.012\,\text{m}^3$

Total volume $= 0.369 + 0.012\,\text{m}^3$

$= 0.381\,\text{m}^3$

Mass = volume × density

$= 0.381 \times 2\,400\,\text{kg}$

$= \underline{914.4\,\text{kg}}$

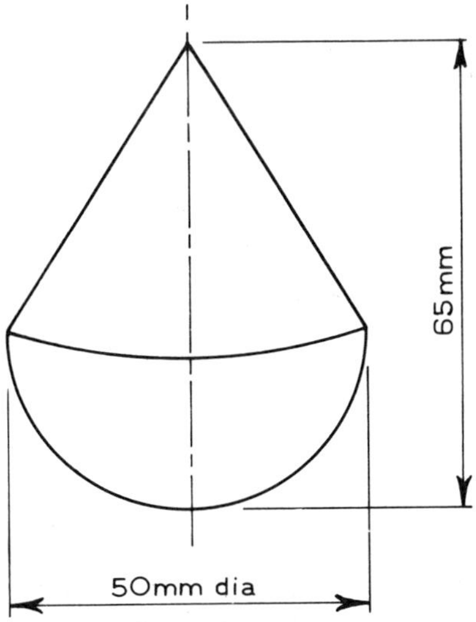

Figure 125

Example A plumb bob made of lead comprises a hemisphere surmounted by a cone. The maximum diameter is 50 mm and the total length 65 mm. Find its volume.

Figure 125 shows the bob.

Volume of hemisphere $= \frac{2}{3}\pi r^3$

Volume of cone $= \frac{1}{3}\pi r^2 h$

Total volume $= \frac{2}{3}\pi r^3 + \frac{1}{3}\pi r^2 h$

$= \frac{1}{3}\pi r^2(2r + h)$

$= \frac{1}{3}\pi \times 25^2(50 + 40)\,\text{mm}^3$

$= \underline{58\,905\,\text{mm}^3}$ or $58.905\,\text{cm}^3$

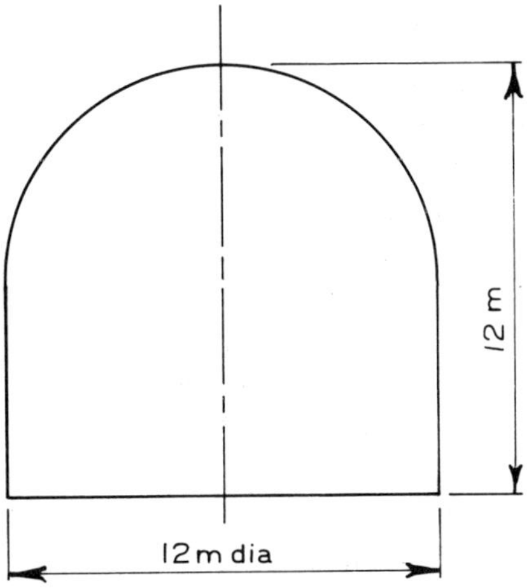

Figure 126

Example A building is circular on plan and has a hemispherical domed roof. Its internal dimensions are 12 m diameter and 12 m maximum height. Find the volume of space enclosed.

Figure 126 shows the building's section.

Volume of cylindrical portion $= \pi r^2 h$

$= \pi \times 6^2 \times 6\,\text{m}^3$

$= 216\pi\,\text{m}^3$

Volume of dome (hemisphere) $= \frac{2}{3}\pi r^3$

$= \frac{2}{3}\pi \times 6^3\,\text{m}^3$

$= 144\pi\,\text{m}^3$

Total volume $= 216\pi + 144\pi$

$= 360\pi\,\text{m}^3$

$= \underline{1\,131\,\text{m}^3}$

Exercise 18b

1. A cylindrical column has a diameter of 750 mm and is 2.5 m high. Find its curved surface area and its volume.

2. A brass cylinder 30 mm in diameter and 45 mm long is to be chromium plated. How many square centimetres of plating are required?

3. Figure 127 shows a steel vee-block. Find its mass, given that the density of steel is 7.8 g/cm^3.

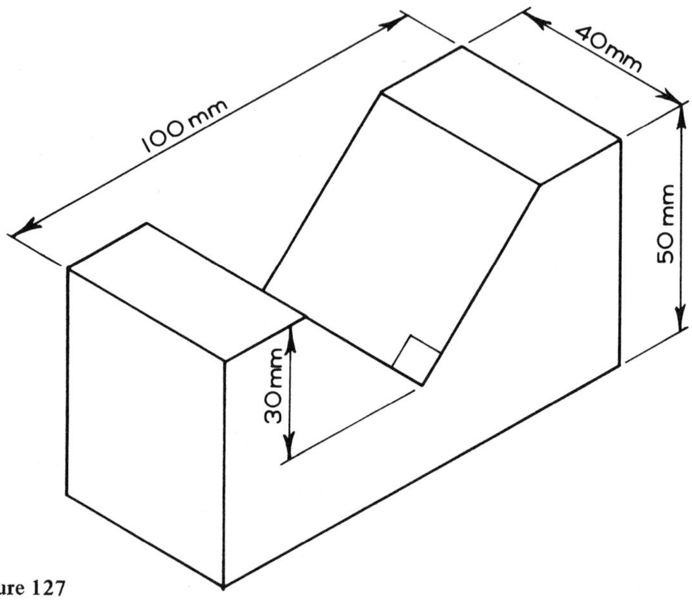

Figure 127

4. A 150 mm concrete cube is immersed in a 1 m diameter cylindrical container holding water to a depth of 300 mm. By how far will the water level rise?

5. The great pyramid in Egypt stands on a 229 m square base and has a maximum height of 144 m. Calculate its volume.

6. A shaft is 12 m deep and has a diameter of 1.75 m for the top quarter of its depth and 1.5 m for the remainder. How much water will it contain when filled to 0.8 of its depth?

7. A storage building is cylindrical up to eaves level and has a conical roof. Its internal diameter is 20 m, the height to eaves level 15 m and the maximum height 17 m. Find its capacity in m^3.

8. A piece of lead pipe 350 mm long having an internal diameter of 40 mm and a wall thickness of 5 mm is melted down and recast into a hemisphere. What is the diameter of the hemisphere? What will be its mass if lead has a density of 11.4 g/cm^3.

9. What proportion of a cubical box is filled by a sphere that fits exactly into it?

10. Two equal cones placed base to base fit exactly into a hollow sphere. What proportion of the sphere remains empty?

11. A swimming bath, rectangular on plan, is 1.5 m deep at the shallow end and 3 m deep at the other end. The bottom has an even gradient and the plan dimensions are 30 m × 12 m. How many litres of water does it contain?

SUMMARY OF F

Angles:

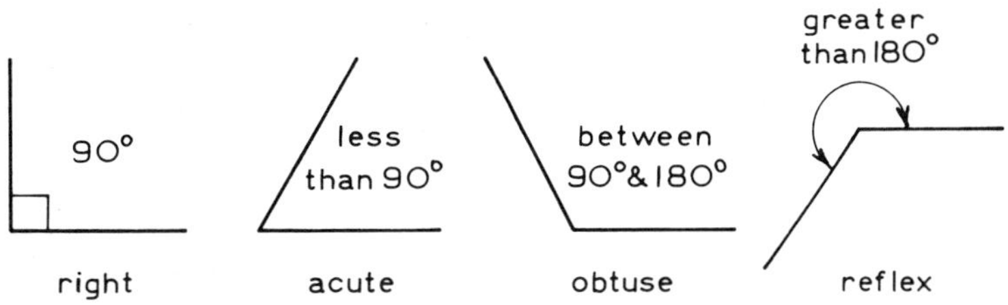

Figure 128

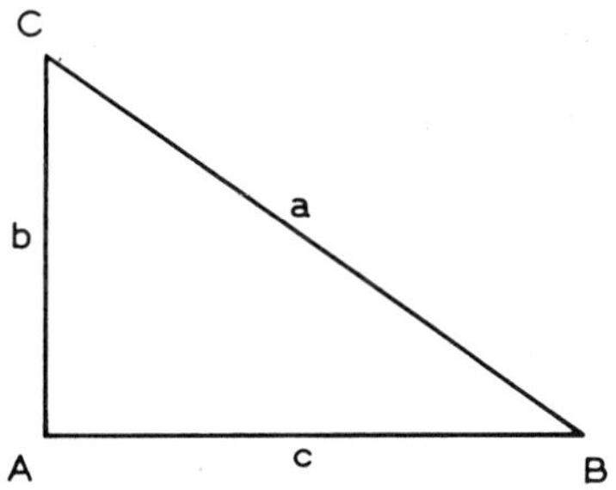

Figure 129

Complementary angles: $\angle A + \angle B = 90°$
Supplementary angles: $\angle A + \angle B = 180°$
Pythagoras (Figure 129):

$$a^2 = b^2 + c^2$$
$$b^2 = a^2 - c^2$$
$$c^2 = a^2 - b^2$$

Circumference of circle: $C = 2\pi r$ (or πd)
Radian measure: 2π radians $= 360°$
Sector: arc $= r\theta$

Areas of plane figures:

Square (length of side s)	s^2
Rectangle (sides l and b)	lb
Triangle (base b, perpendicular height h)	$\frac{1}{2}bh$
Parallelogram (one side b, perpendicular distance to opposite side h)	bh
Trapezium (parallel sides a and b, perpendicular distance between h)	$\frac{1}{2}h(a + b)$
Circle (radius r or diameter d)	πr^2 (or $\frac{1}{4}\pi d^2$)

Volumes of solids:

Prism	Cross-sectional area × length
Pyramid	$\frac{1}{3}$ base area × height
Cylinder	$\pi r^2 h$ (or $\frac{1}{4}\pi d^2 h$)
Cone	$\frac{1}{3}\pi r^2 h$ (or $\frac{1}{12}\pi d^2 h$)
Sphere	$\frac{4}{3}\pi r^3$ (or $\frac{1}{6}\pi d^3$)

Surface area of cylinder:

Curved surface $S = 2\pi rh$
Total surface $S_t = 2\pi r(h + r)$

REVISION EXERCISE F

1. In a rhombus ABCD, AB = 60 mm and $\angle DAB = 60°$. What is the length of the longer diagonal, and what is the value of $\angle DBC$?

2. Two parallel lines, PQ and RS, are intersected by a transverse line at A and B respectively. Sketch the lines and mark a pair of alternate angles (x), a pair of corresponding angles (y) and a pair of supplementary angles (z).

3. Two straight lines, AB and CD, are cut by a transverse line EF at X and Y respectively. $\angle AXE = 125°$ and $\angle DYX = 55°$. Show that AB and CD are parallel.

4. What are the complements and the supplements of the following angles:

a) $15° 30'$ b) $27° 45'$ c) $81° 20'$ d) $45°$

5. Name the type of triangle that has two of its angles less than 45° but not equal. In a triangle ABC, $\angle A = 100°$ and the angle opposite the side b is 50°. What is the third angle, and which is the shortest side?

6. The exterior angle formed when the side AB of an isosceles triangle is produced to a point D is 45°. Find the interior angles of the triangle.

7. Find the unknown sides and angles of each of the following right-angled triangles, $\angle A$ being the right angle in each case:

a) $a = 80$ mm, $b = 40$ mm, $\angle C = 60°$
b) $b = 84$ mm, $c = 73$ mm, $\angle B = 49°$
c) $b = c = 50$ mm

8. Find the side of the square that has a diagonal of length 155 mm.

9. A guy wire stretches from the top point of a 20 m high vertical flag pole to a point on the horizontal ground 8 m from the foot of the pole. How long is the wire?

10. A lean-to roof has a span of 4.5 m and a rise of 2 m. What is the length of the roof slope?

11. Construct the triangles:

a) $a = 83$ mm, $b = 75$ mm, $\angle C = 97°$
b) $\angle A = 90°$, $b = 54$ mm, $c = 85$ mm
c) $\angle B = 50°$, $\angle C = 75°$, $a = 110$ mm

12. A straight wire of length 1 m is bent to the shape of a circle. What is its diameter?

13. A circular loop of wire having a diameter of 500 mm is re-shaped to form an equilateral triangle. What is the size of the triangle?

14. Two chords of the same circle, PQ and RS, meet when produced at a point X outside the circle. $\angle PQR = 50°$ and $\angle RXQ = 35°$. Find the angles PSR and XPS.

15. A tangent to a circle of 5 m diameter is drawn from a point 10 m from its centre. What is the length of the tangent?

16. Convert the following angles to radians:

a) $52° 45'$ b) $89° 15'$ c) $160°$

17. Convert to degrees and minutes:

a) 1.5 radians b) 3.8 radians c) 5.75 radians

18. Express the following angles in radians in terms of π:

a) $12°$ b) $18°$ c) $120°$

19. The curve of a footpath is on a 4 m radius, and the length of the arc it forms is 4.5 m. Through what angle does it turn: a) in radians, b) in degrees and minutes?

20. A sheet-metal balance weight is in the shape of a sector of a circle having an angle of 42° between its radii, which are 110 mm long. What is the weight's perimeter?

21. A trapezium has an area equal to that of a circle of radius 1.75 m. What is the distance between its parallel sides if their lengths are 4.5 m and 6.2 m?

22. A window opening consists of a 1.25 × 1.75 m rectangle surmounted by a semicircle. Find its area and perimeter.

23. What are the cross-sectional area and perimeter of a length of 25 mm quadrant moulding?

24. Find the external surface of an open-ended cylindrical container 450 mm in diameter and 300 mm deep. What is its capacity in litres? Neglect its wall thickness.

25. A cylindrical steel weight of diameter 150 mm and length 200 mm is dropped into a rectangular tank 500 mm square on plan, containing liquid to a depth of 350 mm. By how far will the liquid level rise?

26. The slant height of a cone is 50 mm and the diameter of its base is 60 mm. Find its height, its volume and its mass if the material of which it is made has a density of 2.6 g/cm^3.

27. A solid, composed of a hemisphere surmounted by a cone, has a diameter of 80 mm and an overall length of 100 mm. If it fits exactly into a rectangular box 80 mm square by 100 mm high, how many millilitres of water must be poured in to fill the remaining space?

G Trigonometry

This chapter deals with:
Sine, cosine and tangent ratios of angles
Use of trigonometrical tables
Trigonometrical ratios of the common angles
Solution of right-angled triangles
Sines and cosines of complementary angles
Graphs of sines and cosines

19 THE BASIC TRIGONOMETRICAL RATIOS

19.1 Right-angled triangles

Trigonometry is the special mathematics of *triangles*, dealing with the relationships between their sides and angles. We saw in section 16.2 that a right-angled triangle is one containing an angle of 90° and that its longest side, the *hypotenuse*, is the side *opposite the right angle*. The shortest side is the one opposite the smallest angle. The other two angles of a right-angled triangle are complementary angles.

It is advisable to make sketches of the available data when solving a problem in trigonometry. Although these need not be drawn to scale they should be in good proportion, otherwise the visual interpretation may lead to confusion. This means that the angles should be approximately correct. Reference to section 16.8 will remind you of construction methods. Further useful examples are included in Exercise 19a.

19.2 Sine, cosine and tangent

Certain relationships between the sides and angles of right-angled triangles are expressed as the *basic trigonometrical ratios* of the angles. These ratios are named *sine, cosine* and *tangent* (abbreviated to sin, cos and tan).

In the triangle ABC (Figure 130):

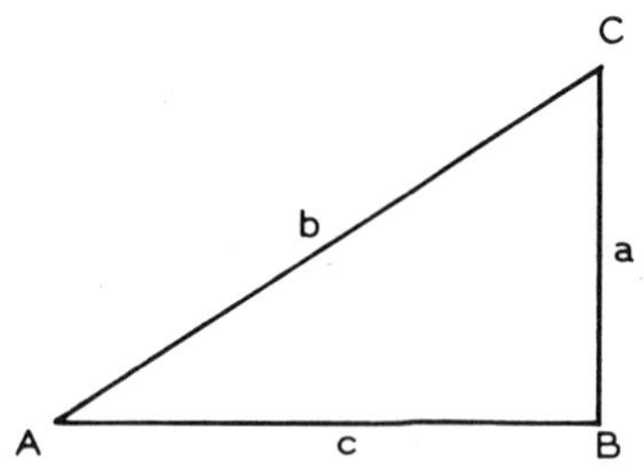

Figure 130

$$\sin A = \frac{a}{b} \quad \left(\frac{\text{side opposite the angle}}{\text{hypotenuse}}\right)$$

$$\cos A = \frac{c}{b} \quad \left(\frac{\text{side adjacent to the angle}}{\text{hypotenuse}}\right)$$

$$\tan A = \frac{a}{c} \quad \left(\frac{\text{side opposite the angle}}{\text{side adjacent to the angle}}\right)$$

Given any one of these basic ratios for an angle, we may construct the angle, as shown in the following examples.

Example Construct the angle that has a tangent of $\frac{3}{5}$.

Since the tangent ratio is *side opposite* to *side adjacent* it involves the two sides containing the right angle. Figure 131 shows the triangle, BC/AB being in the ratio 3 : 5. Measurement of angle A shows it to be 31° (to the nearest degree). Note that the *size* of the triangle is irrelevant, providing the ratio BC : AB is the same.

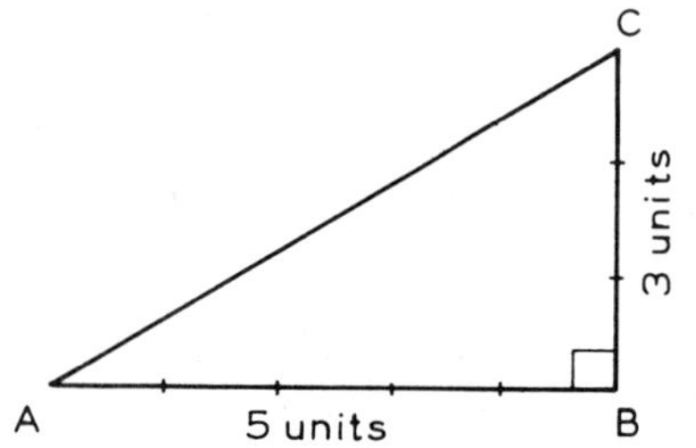

Figure 131

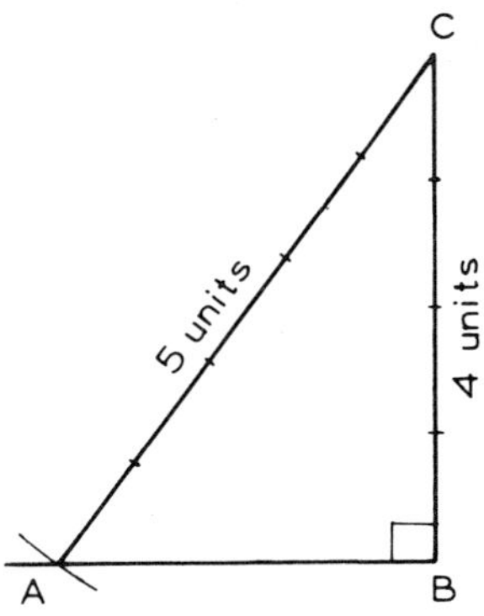

Figure 132

Example By constructing a triangle, find the angle whose sine is 0.8. As a fraction $0.8 = \frac{4}{5}$.

The sine involves the *side opposite* and the *hypotenuse*. In Figure 132 triangle ABC is constructed by drawing the arms of the right angle, making BC = 4 units, and swinging an arc of radius 5 units to cut the other arm at A. Thus $\sin A = \frac{4}{5}$. Measurement of angle A shows it to be 53° (to the nearest degree).

Conversely, given the value of an angle, it is possible to construct a triangle from which the trigonometrical ratios of the angle may be determined.

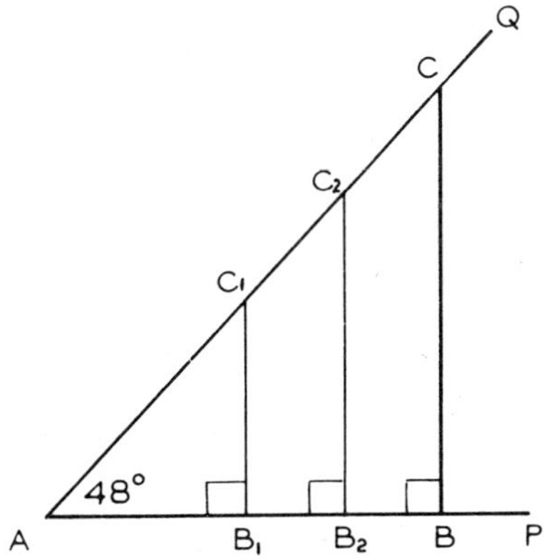

Figure 133

Example Construct a triangle from which to evaluate the sine, cosine and tangent of 48°.

Refer to Figure 133. Using a protractor, draw the lines AP and AQ containing the angle 48°. At any point B on AP erect a perpendicular to meet AQ at C, forming the right-angled triangle ABC. Measure AB, BC and AC.

$$\sin 48° = \frac{BC}{AC} \quad (= 0.74)$$

$$\cos 48° = \frac{AB}{AC} \quad (= 0.67)$$

$$\tan 48° = \frac{BC}{AB} \quad (= 1.11)$$

Note. Wherever the point B is chosen the result will be the same, since all such triangles as AB_1C_1, AB_2C_2 etc. are *similar triangles.*

Exercise 19a

By constructing suitable right-angled triangles, find (to the nearest degree) the angles A having the following trigonometrical ratios:

1. $\tan A = 0.7$ 2. $\sin A = \frac{3}{5}$ 3. $\tan A = \frac{4}{11}$

4. $\cos A = 0.3$ 5. $\cos A = 0.625$ 6. $\sin A = \frac{2}{7}$

Use suitable right-angled triangles to find the values (to three decimal places) of the sines, cosines and tangents of the following angles:

7. 20° 8. 40° 9. 50° 10. 65°

19.3 Trigonometrical ratios from mathematical tables

Values of sines, cosines and tangents for acute angles may be found in your mathematical tables. The trigonometrical tables are read in a similar way to the logarithm tables.

NATURAL SINES

Degrees	0′ 0°.0	6′ 0°.1	12′ 0°.2	18′ 0°.3	24′ 0°.4	30′ 0°.5	36′ 0°.6	42′ 0°.7	48′ 0°.8	54′ 0°.9	Mean Differences 1	2	3	4	5
25	0.4226	4242	4258	4274	4289	4305	4321	4337	4352	4368	3	5	8	11	13
26	0.4384	4399	4415	4431	4446	4462	4478	4493	4509	4524	3	5	8	10	13
27	0.4540	4555	4571	4586	4602	4617	4633	4648	4664	4679	3	5	8	10	13
28	0.4695	4710	4726	4741	4756	4772	4787	4802	4818	4833	3	5	8	10	13
29	0.4848	4863	4879	4894	4909	4924	4939	4955	4970	4985	3	5	8	10	13

From the extract from the table of sines, you will see that for the sine of 28° 32′ we read across the 28° line to the 30′ column, where we find 0.4772. The remainder of the required value is in the 2′ column of the mean differences, where you will find a 5. This is actually 0.0005. Adding this to the 0.4772 gives 0.4777.

i.e. sin 28° 32′ = 0.4777

The next extract is from the tangent tables. To obtain the tangent of 68° 58′ we read across the 68° line to the 54′ column, where we find

NATURAL TANGENTS

Degrees	0′	6′	12′	18′	24′	30′	36′	42′	48′	54′	Mean Differences				
	0°.0	0°.1	0°.2	0°.3	0°.4	0°.5	0°.6	0°.7	0°.8	0°.9	1	2	3	4	5
65	2.1445	1543	1642	1742	1842	1943	2045	2148	2251	2355	17	34	51	68	85
66	2.2460	2566	2673	2781	2889	2998	3109	3220	3332	3445	18	37	55	73	92
67	2.3559	3673	3789	3906	4023	4142	4262	4383	4504	4627	20	40	60	79	99
68	2.4751	4876	5002	5129	5257	5386	5517	5649	5782	5916	22	43	65	87	108
69	2.6051	6187	6325	6464	6605	6746	6889	7034	7179	7326	24	47	71	95	119

2.5916. We now read on to the 4′ column in the same line, where we find 87 (actually 0.0087). Adding this to 2.5916 gives us 2.6003.

i.e. tan 68° 58′ = 2.6003

Note that the tangents of angles up to 45° are less than 1, tan 45° is 1 and tangents of angles greater than 45° are greater than 1.

Now let us look at the extract from the cosine tables. This is different from the sine and tangent tables, in so far as *the greater the angle the smaller its cosine*. For this reason the adjusting values in the 'difference' columns on the right must be *subtracted* instead of added.

NATURAL COSINES

(Numbers in difference columns to be subtracted, not added.)

Degrees	0′	6′	12′	18′	24′	30′	36′	42′	48′	54′	Mean Differences				
	0°.0	0°.1	0°.2	0°.3	0°.4	0°.5	0°.6	0°.7	0°.8	0°.9	1	2	3	4	5
35	0.8192	8181	8171	8161	8151	8141	8131	8121	8111	8100	2	3	5	7	8
36	0.8090	8080	8070	8059	8049	8039	8028	8018	8007	7997	2	3	5	7	9
37	0.7986	7976	7965	7955	7944	7934	7923	7912	7902	7891	2	4	5	7	9
38	0.7880	7869	7859	7848	7837	7826	7815	7804	7793	7782	2	4	5	7	9
39	0.7771	7760	7749	7738	7727	7716	7705	7694	7683	7672	2	4	6	7	9

Suppose we want the cosine of 37° 26′. We read along the 37° line to the 24′ column, where we find 0.7944. Under the 2 in the difference columns we find 4 (0.0004). This is *subtracted from* the 0.7944 to give 0.7940.

i.e. cos 37° 26′ = 0.7940

In many books of mathematical tables, 'subtract mean differences' is printed on cosine tables as a reminder. It is a good idea to underline it neatly in red.

19.4 Angles from tables

It is not convenient to make tables to reverse the above processes as we do with logarithms and antilogarithms. To find an angle given its sine, cosine or tangent the tables must be read in reverse.

Refer again to the extract from the sine tables. To find the angle whose sine is 0.4420, proceed as follows. Search the table for the greatest number less than 0.4420. This is 0.4415, and appears in the 12′ column of the 26° line. Another 0.0005 is needed. A 5 appears in the 2′ column of the differences. Add 2′ to the 26° 12′ already determined and we have 26° 14′. Therefore the angle whose sine is 0.4420 is 26° 14′. This may be written:

$$\sin^{-1} 0.4420 = 26^\circ\ 14'$$

which means exactly the same thing.

Suppose we require the angle whose tangent is 2.3127. Search the tangent tables for the greatest number less than 2.3127. It is 2.3109 in the 36′ column of the 66° line (66° 36′). Another 0.0018 is required. You will find 18 in the 1′ column of the differences, and adding this to 66° 36′ we have 66° 37′. Therefore the angle whose tangent is 2.3127 is 66° 37′. This may be written:

$$\tan^{-1} 2.3127 = 66^\circ\ 37'$$

Now the angle whose cosine is 0.7698. Find the least number greater than this value in the cosine tables. It is 0.7705 and appears in the 36′ column of the 39° line (39° 36′). Since the given cosine is *less* than this, the required angle will be *greater*. The difference is 0.0007, and we find a 7 in the 4′ column of the mean differences. Adding this to 39° 36′ gives 39° 40′.

i.e. $\cos^{-1} 0.7698 = 39^\circ\ 40'$

Exercise 19b

Find the following ratios from mathematical tables:

1. tan 24° 30′	2. sin 32° 40′	3. cos 16° 30′
4. sin 45° 50′	5. tan 6° 18′	6. cos 81° 24′
7. cos 13° 08′	8. sin 6° 18′	9. tan 72° 20′

Use tables to find the following angles:

10. $\tan^{-1} 2.5649$	11. $\cos^{-1} 0.0512$	12. $\tan^{-1} 1.0065$
13. $\cos^{-1} 0.9474$	14. $\sin^{-1} 0.6459$	15. $\tan^{-1} 0.3076$
16. $\sin^{-1} 0.8960$	17. $\cos^{-1} 0.8960$	18. $\tan^{-1} 1.7321$

19.5 The common angles

Certain acute angles are known as the *common angles*. They are 30°, 45° and 60°. You will recognise them as your set square angles.

You will recall from section 16.2 that the 45° triangle in Figure 134 is an *isosceles* triangle, having two equal angles and therefore two equal sides. If we make the equal sides AB and AC each 1 unit in length and apply the Theorem of Pythagoras, we have:

$$a^2 = b^2 + c^2$$

$$= 1^2 + 1^2 = 2$$

$$\therefore\ a = \sqrt{2}\ (1.414)$$

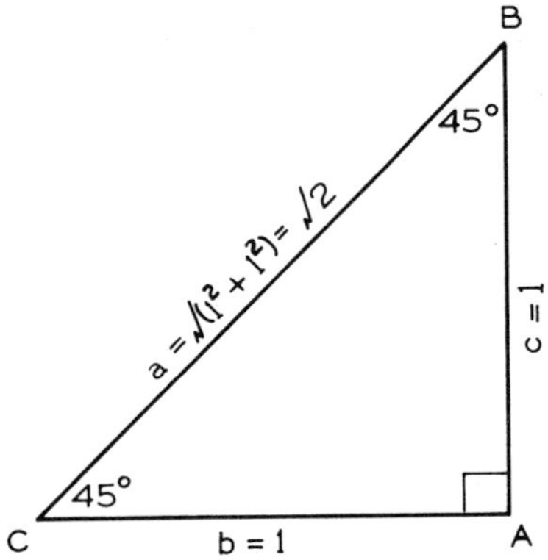

Figure 134

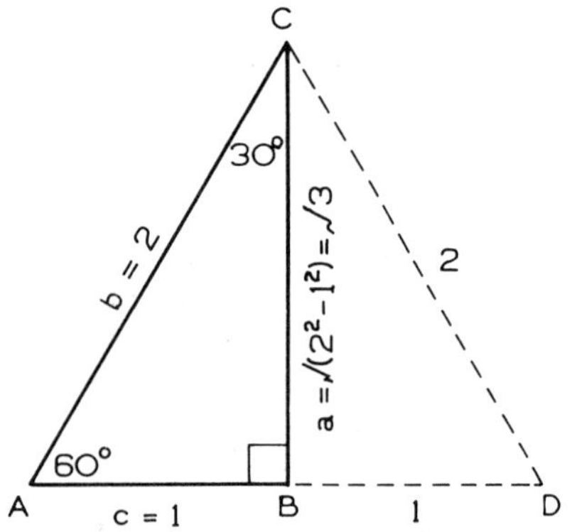

Figure 135

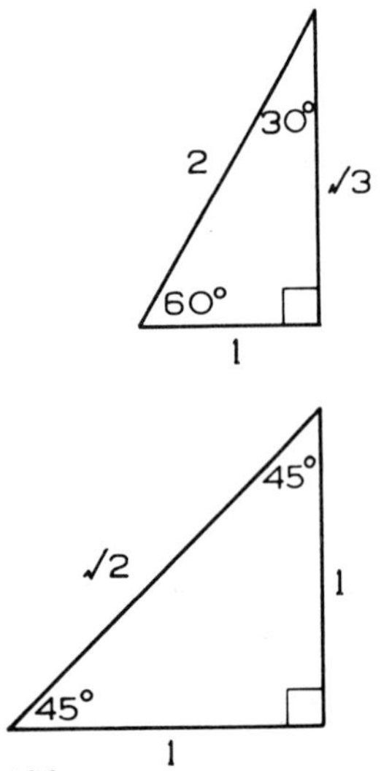

Figure 136

In any 45° right-angled triangle the sides are in the ratio 1 : 1 : $\sqrt{2}$. If we know that the equal sides are 2.5 m each, the hypotenuse is ($2.5 \times \sqrt{2}$) m; if the equal sides are 6.3 m the hypotenuse is ($6.3 \times \sqrt{2}$) m, and so on. Conversely, if we know the length of the hypotenuse of a 45° right-angled triangle, we can obtain the other sides by dividing by $\sqrt{2}$; e.g. if the hypotenuse is 11.6 m the equal sides are each $11.6/\sqrt{2}$ m.

In the triangle ABC in Figure 135, $\angle A = 60°$ and $\angle B = 30°$. If we draw an imaginary triangle, identical but opposite to ABC, we form the isosceles triangle ACD and, since all its angles are 60°, it is an equiangular or equilateral triangle (section 16.2). The 30°-60° triangle ABC is exactly half of it.

If we now make the sides of the equilateral triangle each 2 units in length, AB will be 1 unit and (using Pythagoras' theorem):

$$a^2 = b^2 - c^2$$
$$= 2^2 - 1^2$$
$$= 3$$
$$\therefore \; a = \sqrt{3} \quad (1.732)$$

In any 30°-60° right-angled triangle the sides are in the ratio 1 : 2 : $\sqrt{3}$. Remember that the '2' side is the longest side (the hypotenuse).

Armed with this information we can evaluate the trigonometrical ratios of the common angles without recourse to the tables. From Figure 136:

$$\sin 45° = \frac{1}{\sqrt{2}} \qquad \sin 60° = \frac{\sqrt{3}}{2} \qquad \sin 30° = \frac{1}{2}$$

$$\cos 45° = \frac{1}{\sqrt{2}} \qquad \cos 60° = \frac{1}{2} \qquad \cos 30° = \frac{\sqrt{3}}{2}$$

$$\tan 45° = 1 \qquad \tan 60° = \sqrt{3} \qquad \tan 30° = \frac{1}{\sqrt{3}}$$

$\sqrt{2} = 1.41421\ldots$ etc. and $\sqrt{3} = 1.73205\ldots$ etc. Both are infinite or non-terminating decimals, and such roots are known as *surds*. The ratios of the common angles that have surds as their denominators may be evaluated as follows.

$$\frac{1}{\sqrt{2}} = \frac{1}{\sqrt{2}} \times \frac{\sqrt{2}}{\sqrt{2}} \quad \text{(multiplying numerator and denominator by } \sqrt{2})$$

$$= \frac{\sqrt{2}}{2} = \frac{1.414}{2} = \underline{0.707}$$

The process is called *rationalising the denominator*. Similarly:

$$\frac{1}{\sqrt{3}} = \frac{1}{\sqrt{3}} \times \frac{\sqrt{3}}{\sqrt{3}}$$

$$= \frac{\sqrt{3}}{3} = \frac{1.732}{3} = \underline{0.577}$$

Tabulating the ratios of the common angles we now have:

Angle	30°	45°	60°
sin	$\frac{1}{2}$ or 0.5	$\frac{1}{\sqrt{2}}$ or 0.707	$\frac{\sqrt{3}}{2}$ or 0.866
cos	$\frac{\sqrt{3}}{2}$ or 0.866	$\frac{1}{\sqrt{2}}$ or 0.707	$\frac{1}{2}$ or 0.5
tan	$\frac{1}{\sqrt{3}}$ or 0.577	1	$\sqrt{3}$ or 1.732

19.6 Complementary angles

You will have noticed that:

$\sin 45° = \cos 45°$

$\sin 30° = \cos 60°$

$\sin 60° = \cos 30°$

In other words, *the sine of an angle is equal to the cosine of its complement* and *the cosine of an angle is equal to the sine of its complement.*

Denoting any acute angle by θ, the above statements may be written:

$\sin \theta = \cos (90 - \theta)$

$\cos \theta = \sin (90 - \theta)$

19.7 Solving right-angled triangles

Completing the evaluation of a triangle's properties from given, limited, information is known as *solving the triangle*. Study the following example carefully.

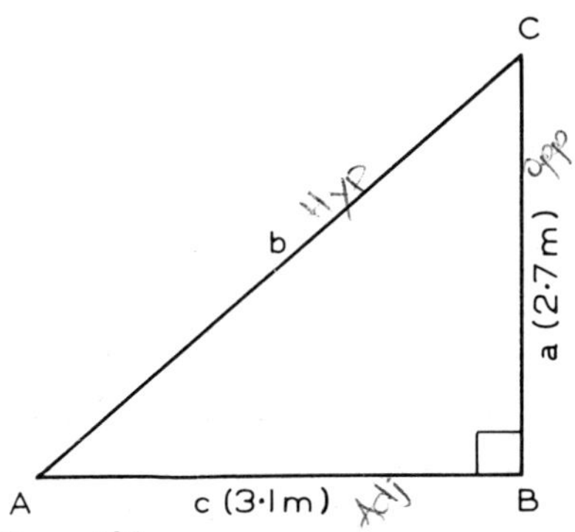

Figure 137

Example Solve the triangle $a = 2.7$ m, $c = 3.1$ m, $\angle B = 90°$

First sketch the triangle (see section 19.1) as in Figure 137. To find side b:

$$b^2 = a^2 + c^2$$

$$= 2.7^2 + 3.1^2$$

$$= 7.29 + 9.61 = 16.9$$

$$b = \sqrt{16.9} = \underline{4.11 \text{ m}}$$

To find $\angle A$: $\tan A = a/c$

$$= 2.7/3.1 = 0.8710$$

$$\angle A = \underline{41° 1'}$$

To find $\angle C$: $\angle A$ and $\angle C$ are complementary

$$\angle C = (90° - \angle A)$$

$$= (90° - 41° 1') = \underline{48° 59'}$$

To find the area: Area = $\frac{1}{2}$ base × height

$$ABC = \tfrac{1}{2} \times 3.1 \times 2.7 \text{ m}^2$$

$$= \underline{4.185 \text{ m}^2}$$

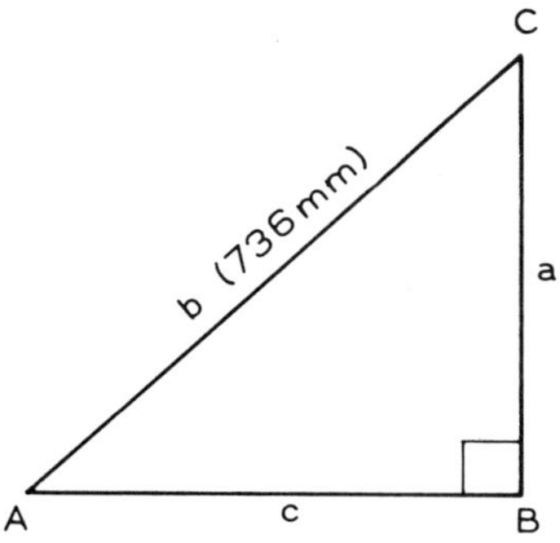

Figure 138

Example Solve the triangle $\angle B = 90°$, $\angle A = 41° 30'$, $b = 736$ mm.

Sketch the triangle (Figure 138) To find $\angle C$:

$\angle A$ and $\angle C$ are complementary

$\angle C = (90° - \angle A)$

$= (90° - 41° 30')$

$= \underline{48° 30'}$

To find side a:

$a/b = \sin A$

$a = b \sin A$

$= 736 \sin 41° 30'$

$= \underline{488 \text{ mm}}$ (to nearest mm)

To find side c:

$c^2 = b^2 - a^2$

$= 736^2 - 488^2$

$= 541\,700 - 238\,100$

$= 303\,600$

$c = \sqrt{303\,600} = \underline{551 \text{ mm}}$ (to nearest mm)

To find the area:

Area $= \frac{1}{2}$ base $\times$ height

ABC $= \frac{1}{2} \times 551 \times 488$

$= \underline{134\,400 \text{ mm}^2}$ (to 4 significant figures)

This could alternatively be expressed as 1.344×10^5 in standard form.

Many problems in surveying and setting out are solved by these methods. The following examples are typical, introducing and defining some of the terms used.

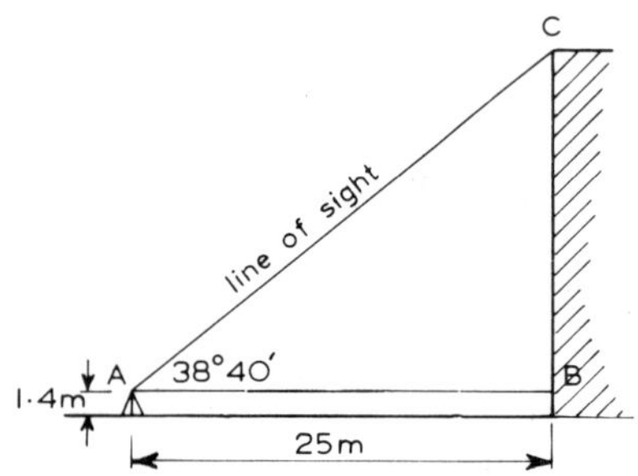

Figure 139

Example The angle of elevation of the topmost point of a building, viewed from a point on level ground 25 m away, is 38° 40'. Find the height of the building if the height of the instrument is 1.4 m.

Angle of elevation is the angle made between the line of sight and the horizontal when an elevated object is viewed from a lower level.

In Figure 139, AC is the line of sight from the instrument A to the top of the building C. In triangle ABC:

$BC/25 = \tan 38° 40'$

BC = 25 tan 38° 40′

= 20 m

Height of building = 20 + 1.4 m = 21.4 m

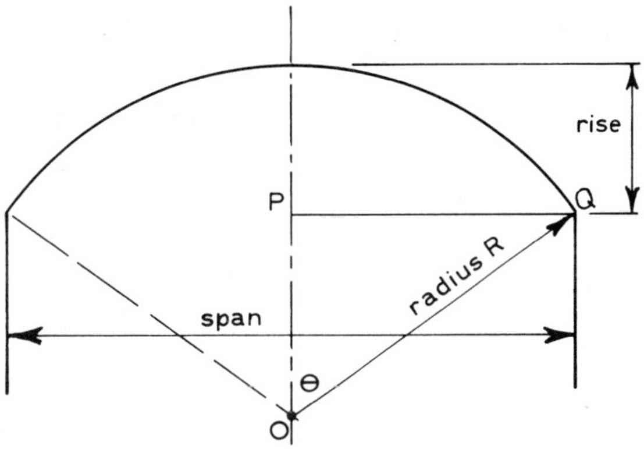

Figure 140

Example A circular segmental arch has a span of 3 m and a rise of 750 mm. Find its radius and the angle it subtends at its centre.

First make a diagram of the arch (Figure 140). Reference back to section 17.1 will remind you of the terms used. In triangle OPQ:

PQ = $\frac{1}{2}$ span = 1.5 m

OP = radius − rise

= $(R - 0.75)$ m

θ is half the required angle

By Pythagoras' theorem:

OQ^2	$= OP^2 + PQ^2$
i.e. R^2	$= (R - 0.75)^2 + 1.5^2$
R^2	$= R^2 - 1.5R + 0.5625 + 2.25$
$1.5R$	$= 2.8125$
R	$= 2.8125/1.5 = \underline{1.875 \text{ m}}$
$\sin\theta$	$= PQ/OQ$
	$= 1.5/1.875$
θ	$= 53° 9'$
Angle at centre	$= 2\theta = \underline{106° 18'}$

Example A roof spanning 12 m has one slope pitched at 38° and the other at 50°, as shown in Figure 141. Find the height of the ridge above the eaves and the two rafter lengths.

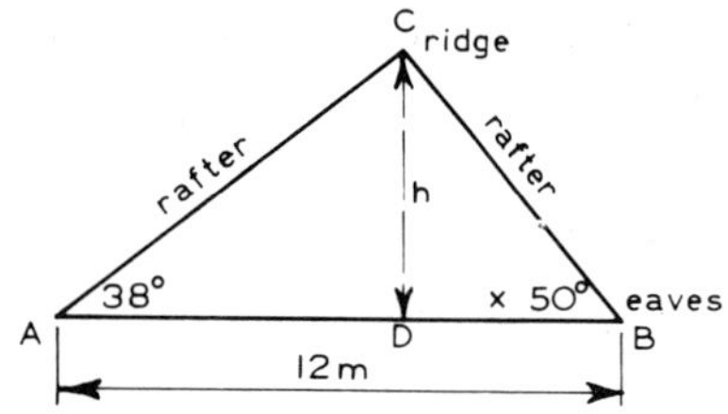

Figure 141

Let the distance DB be x. Then AD = $(12 - x)$. In triangle ADC:

$$h/(12 - x) = \tan 38^\circ$$

$$h = (12 - x)\tan 38^\circ$$

In triangle BDC:

$$h/x = \tan 50^\circ$$

$$h = x \tan 50^\circ$$

Since both equal h:

$$x \tan 50^\circ = (12 - x)\tan 38^\circ$$

$$= 12 \tan 38^\circ - x \tan 38^\circ$$

$$x \tan 50^\circ + x \tan 38^\circ = 12 \tan 38^\circ$$

$$x(\tan 50^\circ + \tan 38^\circ) = 12 \tan 38^\circ$$

$$x = \frac{12 \tan 38^\circ}{\tan 50^\circ + \tan 38^\circ}\text{ m}$$

$$= \frac{12 \times 0.7813}{1.1918 + 0.7813}\text{ m}$$

$$= \frac{12 \times 0.7813}{1.9731}\text{ m}$$

$$= \underline{4.752\text{ m}}$$

$$h = x \tan 50^\circ$$

$$= 4.752 \tan 50^\circ$$

$$= \underline{5.663\text{ m}}\quad\text{(height of eaves)}$$

In triangle ACD:

$$h/\text{AC} = \sin 38^\circ$$

$$\text{AC} = h/\sin 38^\circ$$

$$= 5.66/\sin 38^\circ$$

$$= \underline{9.198\text{ m}}\quad\text{(long rafter)}$$

In triangle BCD:

$$h/\text{BC} = \sin 50^\circ$$

$$\text{BC} = h/\sin 50^\circ$$

$$= 5.663/\sin 50^\circ$$

$$= \underline{7.393\text{ m}}\quad\text{(short rafter)}$$

Example Observations are made from two points P and Q, 250 m apart on a survey line, of a third point R distant from the line. The angles PQR and QPR are measured as 42° 30′ and 65° 20′ respectively. Find the distances PR and QR and the shortest distance from point R to the survey line.

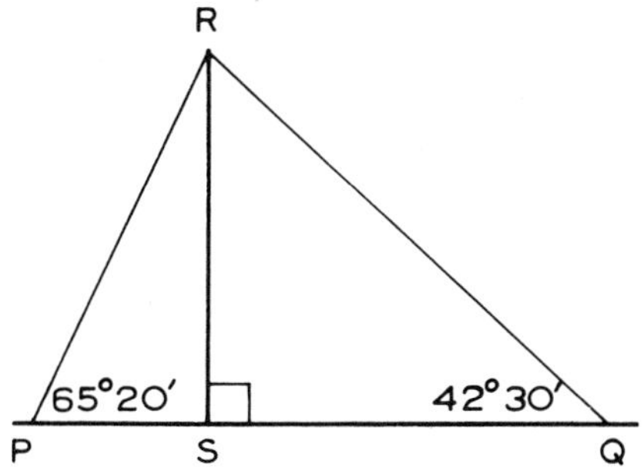

Figure 142

Figure 142 shows the problem. The shortest distance from point R to the line PQ is the perpendicular distance RS.

Let distance PS be x.

$$RS/x = \tan 65° \, 20'$$

$$RS = x \tan 65° \, 20'$$

$$RS/(250 - x) = \tan 42° \, 30'$$

$$RS = (250 - x) \tan 42° \, 30'$$

Since both equal RS:

$$x \tan 65° \, 20' = (250 - x) \tan 42° \, 30'$$

$$x \tan 65° \, 20' = 250 \tan 42° \, 30' - x \tan 42° \, 30'$$

$$x \tan 65° \, 20' + x \tan 42° \, 30' = 250 \tan 42° \, 30'$$

$$x (\tan 65° \, 20' + \tan 42° \, 30') = 250 \tan 42° \, 30'$$

$$x = \frac{250 \tan 42° \, 30'}{\tan 65° \, 20' + \tan 42° \, 30'}$$

$$= \frac{250 \tan 42° \, 30'}{2.1776 + 0.9163}$$

$$= \frac{250 \tan 42° \, 30'}{3.0939}$$

$$= \underline{74.03 \text{ m}}$$

$$RS = x \tan 65° \, 20'$$

$$= 74.03 \tan 65° \, 20'$$

$$= \underline{161.2 \text{ m}}$$

In triangle PRS:

$$PR^2 = PS^2 + RS^2$$

$$= 74.03^2 + 161.2^2$$

$$= 5\,480 + 25\,990$$

$$= 31\,470$$

$$PR = \sqrt{31\,470} = \underline{177.4 \text{ m}}$$

In triangle QRS:

$$QR^2 = QS^2 + RS^2$$

$$= (250 - 74.03)^2 + 161.2^2$$

$$= 175.97^2 + 161.2^2$$

$$= 30\,980 + 25\,990$$

$$= 56\,970$$

$$QR = \sqrt{56\,970} = \underline{238.6 \text{ m}}$$

You may have noticed that the last two problems are similar but have been solved by different methods. Whereas in the roof problem the rafter lengths were found using the sine ratios of the angles, in the survey problem the distances PR and QR were found using the Theorem of Pythagoras. Both methods are equally acceptable.

Exercise 19c Solve the triangles:

1. $\angle B = 90°, c = 83$ mm, $a = 60$ mm

2. $\angle B = 90°, \angle A = 25° 10', c = 150$ mm

3. $\angle A = 50° 27', \angle C = 90°, b = 3.65$ mm

4. $\angle C = 90°, a = 349$ mm, $c = 512$ mm

5. The angle of elevation of the top of a tower from a point 8.5 m away on level ground is 41° 30′. Find the height of the tower if the observation is made from an instrument 1.5 m above ground level.

6. The long sides of a rectangular plot are 45 m, and the diagonal makes an angle of 37° 40′ with them. Calculate the area of the plot.

7. The foot of a ladder is 2.5 m away from a vertical wall on level ground. The angle between the ladder and the ground is 65° 30′. If the foot is moved 200 mm farther away from the wall, how far will the top be lowered?

8. A mast is supported by a guy wire 52 m long, which makes an angle of 30° with the mast. How tall is the mast, and how far away from its foot is the guy wire anchored?

20 GRAPHS OF TRIGONOMETRICAL RATIOS

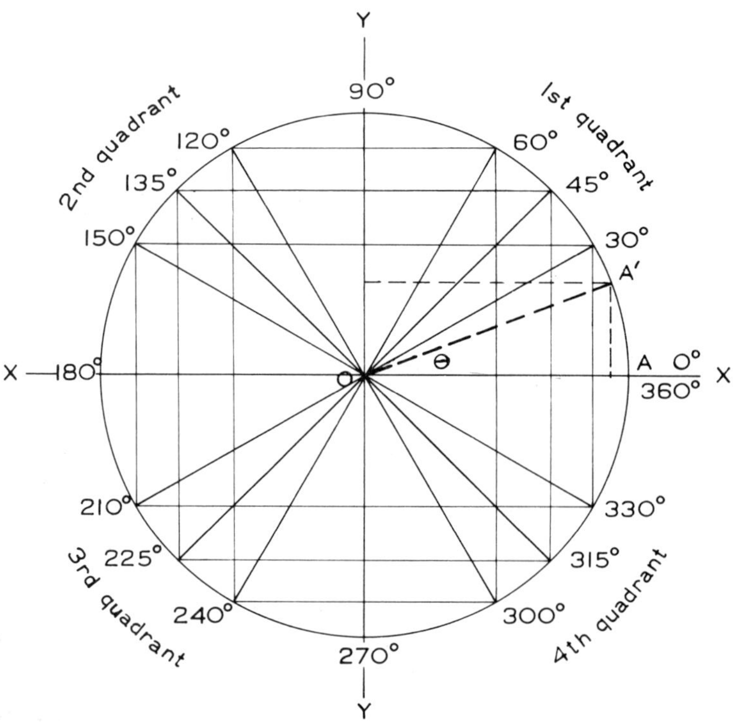

Figure 143

20.1 The sine curve

Consider the line OA in Figure 143, rotating from an initial position at 0° through the four quadrants. For any position of OA′ the sine of the angle swept out from 0° is given by:

$$\sin\theta = \frac{\text{height of A}'}{\text{OA}}$$

If we make OA unit length:

$\sin\theta$ = height of A′

or vertical displacement of A from XX

By convention, distances above the XX axis are positive and distances below negative.

The diagram shows that in the first quadrant the sines increases from 0 to 1 as the angle increases from 0° to 90°. In the second quadrant the sines decrease from 1 to 0 as the angle increases from 90° to 180°. In the third quadrant the sines decrease from 0 to −1 as the angle increases to 270°, and in the fourth quadrant the sines increase from −1 to 0 at 360°.

Now consider the cosines:

$$\cos\theta = \frac{\text{distance of A}' \text{ from vertical axis}}{\text{OA}}$$

and, as OA is unit length:

$\cos\theta$ = distance of A from vertical axis

or horizontal displacement of A from YY

Distances to the right of YY are considered positive and distances to the left negative.

The diagram shows that as OA moves from 0° to 90° the cosine decreases from 1 to 0. From 90° to 180° it continues to decrease from 0 to −1. In the third quadrant as the angle increases from 180° to 270° the cosine increases from −1 to 0, and in the fourth quadrant it continues to increase to 1 at 360°.

The table below shows the ratios sine and cosine for multiples of 30° and 45° up to 360°, from which their graphs may be plotted. Figure 144 shows the sine graph as a smooth and continuous oscillating curve.

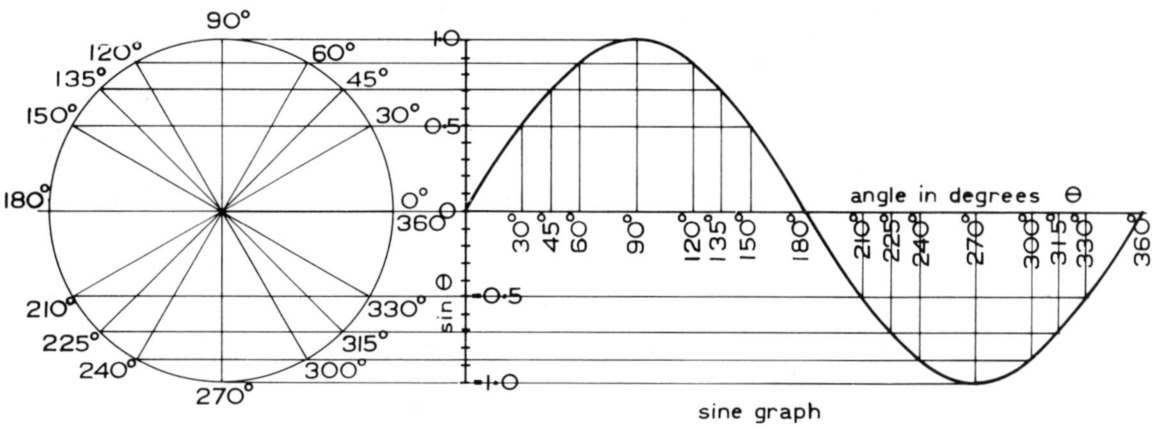

Figure 144

Angle	0°	30°	45°	60°	90°	120°	135°	150°	180°
sine	0	0.5	0.707	0.866	1	0.866	0.707	0.5	0
cosine	1	0.866	0.707	0.5	0	−0.5	−0.707	−0.866	−1

Angle	210°	225°	240°	270°	300°	315°	330°	360°
sine	−0.5	−0.707	−0.866	−1	−0.866	−0.707	−0.5	0
cosine	−0.866	−0.707	−0.5	0	0.5	0.707	0.866	1

Although we have calculated the sine values for the graph, you can see that it may also be constructed geometrically from the positions of the revolving line OA. They are simply projected across to the appropriate ordinate on the graph.

20.2 The cosine curve

A similar construction may be used for the cosine graph, but it must be plotted on a vertical axis since the cosine values are the horizontal displacements of A. This is shown in Figure 145. The horizontal plotting of the cosine graph, as in Figure 146, is the more usual method of presentation.

Figure 146 shows the sine curve and the cosine curve together on the same axes. Comparison of the graphs confirms such basic relationships as:

$\sin \theta = \cos (90 - \theta)$

$\sin \theta$ increases as θ increases for acute angles

$\cos \theta$ decreases as θ increases for acute angles

If you note the point of intersection of the two curves in the first quadrant, you will see that it is on the 45° ordinate,

i.e. $\sin 45° = \cos 45°$

If you read across the 0.5 line, you will see that in the first quadrant it cuts the sine curve at 30° and the cosine curve at 60°,

i.e. $\sin 30° = \cos 60°$

If you find the level 0.866 and read across the first quadrant you will find it cuts the cosine curve at 30° and the sine curve at 60°,

i.e. $\cos 30° = \sin 60°$

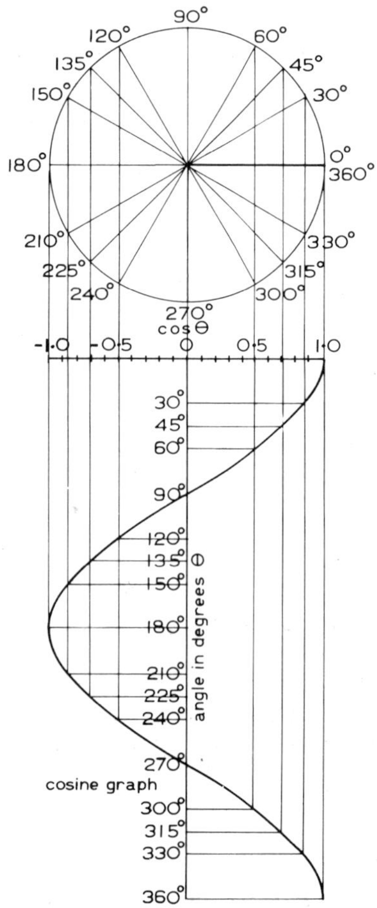

Figure 145

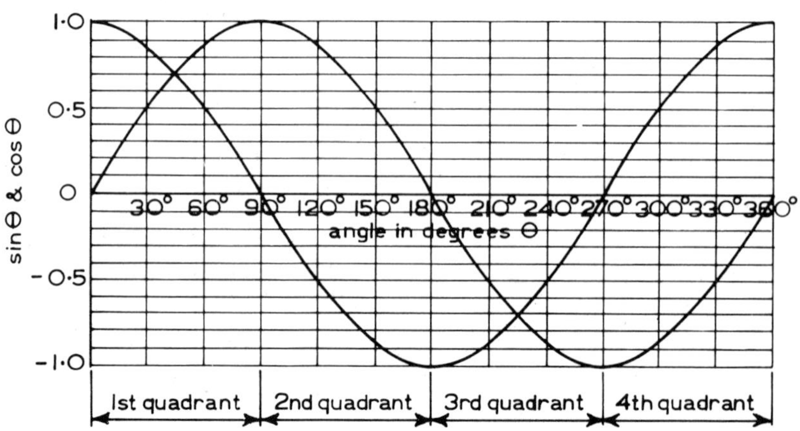

Figure 146

The graph also shows that sines are positive in the first and second quadrants but negative in the third and fourth quadrants. Cosines are positive in the first and fourth quadrants, negative in the second and third quadrants. We shall consider this aspect of the trigonometrical ratios in more detail in Book 2.

Exercise 20

1. Draw up a table of the sines of angles between 0° and 90° in intervals of 10°, and plot a graph of the result. The angles should be plotted on the horizontal axis and the values of their sines on the vertical axis. From the graph read off the values of sin 15°, sin 25°, sin 35°, sin 55°, sin 65°, sin 75° and sin 85°. Check your results from the sine tables.

2. Plot a graph of the sines of angles between 0° and 45° in stages of 5° by projecting the positions of the end A of a revolving line OA as in Figure 144. Use as large a scale as possible. From your graph read off the following values: sin 17°, sin 23°, sin 39°. Check your results against the sine tables.

   Can you now find a way of using the same graph to find the *cosines* of 27° and 31°?

3. Make a table of the cosines of angles between 50° and 80° in stages of 5°, and plot a graph of the results to the largest scale possible. From the curve estimate the values of the cosines of 52°, 71° and 79°.

Use your own version of Figure 146, and the sine and cosine tables, to determine the following ratios. Compare the values obtained by the two methods.

4. sin 175°	5. cos 212°	6. sin 96°
7. cos 312°	8. sin 111°	9. sin 325°
10. cos 106°	11. cos 273°	12. sin 241°

SUMMARY OF G

$$\text{sine} = \frac{\text{opposite}}{\text{hypotenuse}}$$

$$\text{cosine} = \frac{\text{adjacent}}{\text{hypotenuse}}$$

$$\text{tangent} = \frac{\text{opposite}}{\text{adjacent}}$$

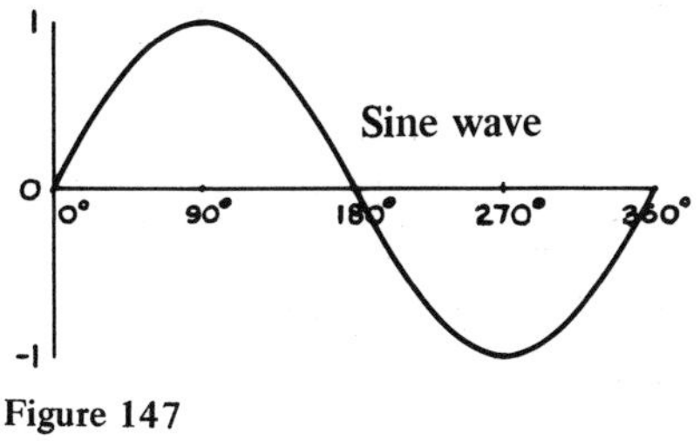

Figure 147

Sine, tangent: the sine and tangent of an acute angle *increase* as the angle *increases*.

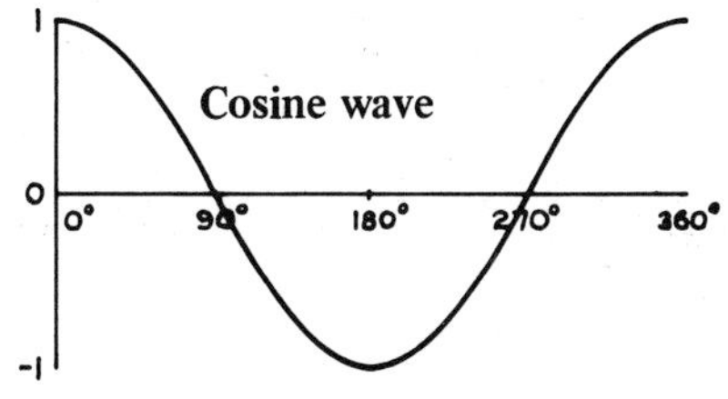

Figure 148

Cosine: the cosine of an acute angle *decreases* as the angle *increases*. Mean differences must be *subtracted*, not added, when using the cosine tables.

$\sin\theta = \cos(90 - \theta)$

$\cos\theta = \sin(90 - \theta)$

$\sin^{-1} x$ means the angle whose sine is x, and similarly for the other ratios.

REVISION EXERCISE G

Make accurate constructions of the following triangles, measure the remaining sides and/or angles, and calculate the three basic trigonometrical ratios of the angles:

1. $\angle B = 90°$, $a = 35$ mm, $c = 55$ mm
2. $\angle A = 35°$, $\angle B = 90°$, $b = 62$ mm
3. $\angle A = 90°$, $b = 27$ mm, $c = 43$ mm
4. $\angle A = 71°$, $\angle C = 90°$, $a = 80$ mm

Obtain the following ratios from mathematical tables:

5. sin 27° 30′   6. tan 53° 08′   7. sin 59° 10′

8. cos 53° 17′   9. tan 68° 10′   10. cos 15° 39′

Construct and measure the following angles:

11. $\tan^{-1} \frac{5}{6}$ 12. $\sin^{-1} \frac{2}{5}$ 13. $\cos^{-1} 0.625$

Use tables to find the following angles:

14. $\tan^{-1} 1.2736$ 15. $\cos^{-1} 0.5430$

16. $\sin^{-1} 0.7002$ 17. $\tan^{-1} 0.3640$

Solve the triangles:

18. $\angle B = 90°$, $c = 8.7$ m, $a = 5.75$ m
19. $\angle C = 90°$, $a = 6.9$ m, $c = 10.3$ m
20. An equilateral triangle has sides of 68 mm. Find its area.
21. The equal sides of an isosceles triangle are 75 mm in length and the angle between them is 39°. Solve the triangle.
22. From the top of a tower the angle of depression of a point 200 m away on level ground is 21° 30′. How high is the tower? (Angle of depression is the angle between the line of sight and the horizontal when an object is viewed from a higher level.)
23. The mid-point of a chord is 55 mm from the centre of the circle, and the chord subtends an angle of 68° at the centre. Find the diameter of the circle and the length of the chord.
24. Complete the table below and plot a graph of the results, the angle in radians being on the x axis.

Angle in radians	$\pi/6$	$\pi/3$	$\pi/2$	$2\pi/3$	$5\pi/6$	π
Cosine of angle						

25. Plot, on the same pair of axes, graphs of $\cos\theta$ and $\tan\theta$ for values of θ between 0° and 45°. From the graph find the angle for which $\cos\theta = \tan\theta$.
26. Plot the graphs of $2\sin\theta$ and $3\cos\theta$ on the same pair of axes at intervals of 20° for values of θ from 0° to 180°. Use the graph to find the angle for which $2\sin\theta = 3\cos\theta$.

Answers to exercises

Exercise 1a	1. 1	2. 3	3. 0	4. −12	
	5. −2	6. 2	7. −6	8. 24	
	9. −45	10. 2	11. −9	12. −0.5	
	13. $2 \times 3 \times 5 \times 7$		14. $3 \times 7 \times 11$		
	15. $2 \times 5 \times 13 \times 13$		16. $2 \times 2 \times 2 \times 2 \times 5 \times 7$		
	17. $3 \times 7 \times 7$		18. 4		
	19. No common factor		20. 15		
	21. 16	22. 6	23. 17	24. 24	
	25. 135	26. 84	27. 90	28. 72	
	29. 42				
Exercise 1b	1. $14\frac{1}{4}$	2. $29\frac{3}{8}$	3. $3\frac{1}{7}$	4. $8\frac{1}{5}$	5. $4\frac{49}{144}$
	6. $3\frac{5}{16}$	7. $\frac{39}{7}$	8. $\frac{58}{5}$	9. $\frac{135}{8}$	10. $\frac{1307}{128}$
	11. $\frac{343}{15}$	12. $\frac{451}{64}$	13. $7\frac{29}{42}$	14. $7\frac{44}{45}$	15. $15\frac{19}{30}$
	16. $6\frac{7}{32}$	17. 77	18. 7.5	19. $3\frac{19}{52}$	20. $11\frac{47}{88}$
	21. $7\frac{6}{7}$	22. 16	23. $3\frac{1}{3}$	24. $\frac{189}{500}$	25. $6\frac{975}{1024}$
Exercise 2a	1. 0.328	2. 0.667	3. 0.625	4. 0.25	
	5. 0.438	6. 0.675	7. 3.778	8. 0.45	
	9. 1.56	10. $\frac{7}{8}$	11. $\frac{1}{5}$	12. $\frac{9}{20}$	
	13. $10\frac{9}{25}$	14. $\frac{1}{8}$	15. $6\frac{1}{40}$	16. $0.8\dot{3}$	
	17. $0.3\dot{8}$	18. $0.\dot{7}0\dot{3}$	19. 87.5%	20. 44%	
	21. 97%	22. 17%	23. 71%	24. 84.5%	
	25. 73.5%	26. 12.5%	27. 27%		
Exercise 2b	1. 209.6	2. 361.8	3. 4.925	4. 3.362	
	5. 19.046	6. 125.372	7. 5.228	8. 10.142	
	9. 305.436	10. 4.713	11. 4.159	12. 0.395	
Exercise 3	1. 128	2. 125	3. 216	4. 0.64	
	5. 1.21	6. 1024	7. 9	8. 7	
	9. 2	10. 3	11. 8	12. 5	
	13. $3^7 = 2\,187$		14. $2^8 = 256$		
	15. $\dfrac{3^3 \times 3^4}{3^5} = 9$		16. $\dfrac{2^3 \times 2^6 \times 2^8}{2^4 \times 2^7} = 64$		

17. $\frac{7^2 \times 7^3 \times 7^1}{7^3} = 343$ 18. $\frac{5^2 \times 5^1}{5^2} = 5$

19. 144 20. 8 21. $\frac{1}{16}$ 22. 729

23. 16 24. 64

Exercise 4a

1. 7.329×10^3 2. 1.49×10^4 3. 3.25×10^5
4. 9.8×10^2 5. 6.3×10^{-3} 6. 2.7×10^{-6}
7. 8×10^{-4} 8. 2.71×10^5 9. 1.49×10^{-2}
10. 8 200 000 11. 790 12. 103 000
13. 25 000 000 14. 0.003 2 15. 0.000 001 8
16. 0.000 65 17. 0.000 010 1 18. 0.000 000 581
19. 1.877×10^6 20. 3.663×10^5 21. 2.65×10^5
22. 5.3×10^4 23. 4.938×10^6 24. $6.385\,7 \times 10^7$

Exercise 4b

1. 101 2. 11010 3. 11
4. 1110 5. 10010 6. 100110
7. 10001 8. 110011 9. 111111
10. 5 11. 25 12. 21
13. 59 14. 87 15. 3
16. 9 17. 27 18. 2
19. 10111 (13 + 10 = 23) 20. 100101 (27 + 6 + 4 = 37)
21. 101000 (29 + 11 = 40) 22. 1011000 (55 + 31 + 2 = 88)
23. 1011 (22 − 11 = 11) 24. 110 (11 − 5 = 6)
25. 10011 (18 − 5 + 6 = 19) 26. 1111 (31 − 11 − 5 = 15)

REVISION EXERCISE A

1. a) 13×7 b) none c) 11×11 d) 9×5, $3 \times 3 \times 5$
2. a) and d)
3. a) $3 \times 3 \times 3$ b) $2 \times 2 \times 3 \times 7$ c) 7×31 d) 3×13
4. a) HCF = 2, LCM = 1 008 b) HCF = 16, LCM = 192
   c) HCF = 5, LCM = 1 400 d) HCF = 3, LCM = 504
5. a) $10\frac{1}{12}$ b) $2\frac{1}{42}$ c) $19\frac{7}{8}$ d) $10\frac{1}{24}$
6. $\frac{6}{17}$, $\frac{12}{37}$, $\frac{17}{53}$, $\frac{8}{25}$
7. a) 0.76 b) 0.048 c) 0.421 875 d) 0.821 5
8. a) $\frac{11}{200}$ b) $\frac{5}{16}$ c) $\frac{1}{16}$ d) $\frac{7}{80}$
9. a) 114.105 b) 52.203 8 c) 50.75 d) −0.131
10. a) $3\frac{1}{16}$ b) $\frac{8}{27}$ c) $\frac{1}{64}$ d) 4
11. a) 11 b) $\frac{3}{8}$ c) 3 d) 5

12.	a) $\sqrt{3}$	b) 1	c) $\frac{1}{25}$	
13.	a) $\frac{1}{4^{\frac{1}{2}}}$	b) $8^{\frac{2}{3}}$	c) 1^1	d) 1^1
14.	a) $\frac{1}{7}$	b) 64	c) 27	
15.	a) 4.372×10^6	b) 6.897×10^3	c) 3.247×10^5	
16.	a) 6 300 000	b) 0.005 95	c) 47 000	
17.	a) $2.914\,6 \times 10^6$	b) 4.652×10^6	c) $4.566\,7 \times 10^5$	
18.	a) 111001	b) 10111	c) 100011	d) 101001
19.	a) 19	b) 25	c) 41	
20.	a) 11010	b) 11001	c) 11010	d) 101000

Exercise 5

1. Two places of decimals
2. Three places of decimals
3. Four significant figures

4. 30	5. 30	6. 16	7. 2
8. 0.5	9. 0.1	10. 315	11. 1 500
12. 0.2	13. 1		

Exercise 6a

1. 6.554	2. 137.8	3. 1.573	4. 778.4
5. 0.680 6	6. 0.009 216	7. 7.2	8. 1.718
9. 10.57	10. 3.782	11. 0.918 7	12. 72.63
13. 2.16	14. 0.045 96	15. 0.318 7	16. 0.101 9
17. 0.009 259	18. 13.07		

Exercise 6b

1. 2.052 7	2. 1.754 6	3. 0.880 5	4. $\bar{1}.091\,3$
5. 0.017 0	6. $\bar{3}.988\,1$	7. 3.516 3	8. 4.059 2
9. 6.587 7	10. 4.682 1	11. 2.243 0	12. 5.815 6

Exercise 6c

1. 69.76	2. 104.6	3. 5.703	4. 0.334 3
5. 0.015 19	6. 4 193	7. 63.8	8. 550.1
9. 32.43	10. 7.771	11. 4.315	12. 40 (39.99)
13. 0.025 63	14. 0.000 286 6	15. 54.03	16. 0.274 6
17. 761	18. 45.16	19. 40.22	20. 0.005 853
21. 24.35	22. 3.457	23. 0.795 0	24. 0.097 68
25. 8.521	26. 328.8	27. 14 940	
28. Rise 173 mm, Going 215 mm		29. 1 266 mm	30. 2.784 mm

Exercise 7a

1. 520.7	2. 16 307.5	3. 1335	4. 3.237
5. 0.049	6. 10.87	7. 153.3	8. 2.245
9. 2 715	10. 92.9	11. 40 250	12. 378

	13. 41.4	14. 159.3	15. 2.91	16. 7.39
	17. 0.024	18. 1.585	19. 113.8	20. 0.577
	21. 1.465	22. 0.0174	23. 0.3181	
Exercise 7b	1. 132.3	2. 0.4225	3. 0.00578	4. 26.5
	5. 107000	6. 1722	7. 2.85	8. 13.11
	9. 9.58	10. 0.2608	11. 0.0728	12. 96.7
	13. 13.3	14. 0.606	15. 0.0781	16. 0.1482
	17. 0.008	18. 0.01212		
Exercise 7c	1. 2.745	3. 12.68	3. 24.9	4. 2.84
Exercise 8	1. 34.283	2. £566.60	3. 1.963	4. 260.18
	5. 14.343	6. 20.706	7. 74.2063	8. 37.92
	9. 0.672	10. 5.28	11. 20.05	12. 3.23
	13. 44.82	14. 3.516	15. 0.7396	16. 12.978
	17. 16129	18. 0.9269	19. 18.36	20. 0.0588
	21. 0.0426	22. 0.0031	23. 154.69 kN	24. 2481.5 kg/m^3
	25. 1.577 kW	26. 11495 kJ		

REVISION EXERCISE B

1. One place of decimals (45.1)
2. Three places of decimals (5.690)
3. Two places of decimals (37.56)

4. 50	5. 1.5	6. 20	7. 250
8. 0.4302	9. 0.01869	10. 27.62 Ω	11. 145.7 m

12. 1) 0.92% 2) 0.315% 3) 0.48%

13. 86.4 kJ	14. 490.5 m	15. 31.269 kN/m^2
16. 7.916 MJ		

Exercise 9

1. $5x + 3y$	2. $11x + 8xy$	3. $30a + 13b - 3c$
4. $3x + 3y + z$	5. $x + 12y - 3z$	6. $-6yz$
7. $18x^{10}$	8. $20x^6y^3$	9. $2a^7b^3$
10. $-35x^3y^4$	11. $3xy^2$	12. $\dfrac{2a^3x}{y^2}$
13. $8x - 4xy$	14. $7ab^2 + 4$	15. $9x^2y^4$
16. $49a^6b^2$	17. $25x^4y^6z^2$	18. $16a^2b^4c^2$
19. $\dfrac{8x^{12}}{27y^6}$	20. a^8x^4	21. $\dfrac{1}{729x^{12}y^6}$
22. $-\dfrac{x^{15}}{32}$	23. $8a + 7$	24. $6xy^2$

Exercise 10

1. $x^2 + 7x + 10$
2. $x^2 - 6x - 7$
3. $2x^2 + 10x + 12$
4. $12x^2 - 25x + 12$
5. $x^3 - 10x^2 + 28x - 16$
6. $4a^4 + 7a^3 - 23a^2 + 14a - 5$
7. $a^6 + 3a^4b^2 + 3a^2b^4 + b^6$
8. $0.5x - 2$
9. $x^6 - x^3y^3 + y^6$
10. $x^2 - 3x + 5$
11. $5a^2 + 2b^2$
12. $x(x^2 + 2x - 1)$
13. $x(ax^2 + bx - c)$
14. $x(x + y)$
15. $9a^2(3a^2 - a + 6)$
16. $19p^2q^2(p^2q - 2 + 3pq^2)$
17. $(x + 1)(x + b)$
18. $(m + 4)(n - p)$
19. $(x - p)(y + q)$
20. $(2p - 1)(x + 4y)$
21. $b(a + c)(ac - b)$
22. $(x + 4y)(2p - 1)$
23. $(x + 7)(x + 3)$
24. $(x + 8)(x + 1)$
25. $(x + 2y)^2$

Exercise 11a

1. 8
2. 4
3. 1
4. 2
5. −15
6. −1
7. $6\frac{2}{3}$
8. 2
9. −151

Exercise 11b

1. 3 m × 5 m
2. 6
3. At $\frac{1}{3}$ distance from pivot to 0.5 kg
4. 1.83 m from left end

Exercise 11c

1. $x = 3,\ y = 2$
2. $x = 1.5,\ y = 0.5$
3. $x = 1,\ y = 2.5$
4. $x = 3,\ y = 10$
5. $x = 4,\ y = 3$
6. $x = 0.5,\ y = 0.1$

Exercise 12

1. $1.44 \times 10^4\ \mathrm{N/m^2}$
2. 76.03 kN
3. $210\ \mathrm{kN/mm^2}$
4. 23.69 m/s
5. 12
6. 6.278 kJ
7. $x = \dfrac{Wl}{AE}$, 0.12 mm
8. $t = \dfrac{1}{\alpha_0}\left(\dfrac{R_t}{R_0} - 1\right)$, 50 °C
9. $r = \sqrt{\dfrac{3V}{\pi h}}$, 300 mm
10. 53 mm

REVISION EXERCISE C

1. $14p + 3q + 3r$
2. $8r + 9s - 4t$
3. $12pq + 10qr - 7rs$
4. $-6x^2 + 4x - 2$
5. $28p^6q^8r^8$
6. $-3a^3b^2c^5$
7. $\dfrac{3a}{bc}$
8. $\dfrac{6x^2}{y} + 4xy - 2x^2y$
9. $\dfrac{1}{9x^2y^2}$
10. $\dfrac{9a^2b^4}{16x^4y^2}$
11. $49p^4q^2r^6$
12. $\dfrac{8}{27}x^6y^9$

13. $x^3 + 9x^2 + 16x - 3$

14. $24 + 164y - 170x$

15. $x^2 + 16x + 64$

16. $6x^2 + 4x - 2$

17. $x^3 - 4x^2 + 5x - 20$

18. $3x^3 + 16x^2y + 20xy^2 - 3y^3$

19. $\frac{1}{4}x^3 + \frac{5}{3}x^2 + \frac{5}{6}x - \frac{1}{9}$

20. $x^5 - x^4y + xy^4 - y^5$

21. $x - 2y$

22. $45xy(x - 5y)$

23. $3abc(1 + 3abc)$

24. $11xy(11x^4 - x^2y + 5y^4)$

25. $(x - y)(x + 3)$

26. $y(x - y)(2x + z)$

27. $(3x - y)(a + 2b)$

28. $(pq - p)(p^2 + q^2)$

29. 2

30. 1.5

31. 2

32. 3

33. $x = 15$, $y = 14$

34. $x = \dfrac{1}{4}$, $y = \dfrac{1}{6}$

35. 3

36. 24

37. 3.3 Ω, 1.1 Ω, 0.2 Ω

38. 40 m × 68 m

39. £20 000

40. 115.5 m

41. 28.67%

42. $R_1 = 23.59$ kN, $R_r = 14.53$ kN

43. 38.2 mm, $d = \sqrt{\dfrac{4F}{\pi f}}$

44. $l = \dfrac{RA}{\rho}$, 1.696 m

45. $w = \dfrac{2R_L - W}{l}$

46. $f_s = \dfrac{F}{\pi dt}$, 247.6 N/mm²

47. $a = \dfrac{v - u}{t}$, 1.042 m/s²

48. 792 mm

Exercise 13a

1. 234 m², 89.7 yd²
2. 255 g, 0.88 oz
3. 2307 kg/m³, 487 lb/ft³
4. 35.5 litres, 7.7 gal
5. 116 N/mm², 9.7 tonf/in²
6. 0.4 kJ/kg °C, 0.6 Btu/lb °F

Exercise 13b

1. 348.75 kg, 8.4 m
2. 6.4 amps, 6 volts
3.

°F	10	50	86	150	176	248
°C	−12	10	30	65.5	80	120

4. 2.7 mm, 4.1 mm, No. 9

Exercise 13c

1. 9 square units
2. 7 units
3. a) at the origin b) −1 and 1 c) $-1\frac{1}{3}$ and 4 d) $\frac{1}{2}$ and −1
4. Parallelogram
5. a) $\frac{4}{3}$ b) 3 c) $\frac{5}{2}$ d) $\frac{4}{3}$
6. 3 and −4.5
7. $2\frac{1}{3}$ and $1\frac{3}{4}$

REVISION EXERCISE D

1. 75.94 J, 250 ft lbf
2. 26.1 kW, 13.4 hp
3. 3.6 kN/m², 177.5 lbf/ft²
4. a) 65 N b) 57 N

5. a) 158 mm b) 0.243 kg c) 0.357 kg
6. (6, 7) and (–3, –2)
8. (1, 0)
9. $x = 2.6$, $y = 1.2$
10. (2.25, 0.97)
11. 3 and –2
12. 0.2 and –0.5, 1.55 and –3.4, 2.67 and 4

Exercise 14a

1.

0 to 9	2	50 to 59	15
10 to 19	4	60 to 69	12
20 to 29	7	70 to 79	8
30 to 39	9	80 to 89	5
40 to 49	15	90 to 100	3

2. Discrete.

3 points	1	8 points	11
4 points	2	9 points	8
5 points	3	10 points	6
6 points	5	11 points	3
7 points	10	12 points	1

3. Continuous.

3 to 3.19	1
3.2 to 3.39	2
3.4 to 3.59	5
3.6 to 3.79	9
3.8 to 3.99	14
4.0 to 4.19	11
4.2 to 4.39	4
4.4 to 4.59	3
4.6 to 4.79	1

4.

Bungalows	18.75%
Maisonnettes	8.75%
Semi-detached houses	31.875%
Detached houses	18.125%
Terraced houses	22.5%

5.

Walls	33.6°
Floors	96°
Chimneys	57.6°
Sills	115.2°
Roofs	57.6°

6. Symbol such as  representing 100 men

Exercise 14b

3.

Strength in N/mm²	*Frequency*
14 to 16.9	1
17 to 19.9	3
20 to 22.9	6
23 to 25.9	11
26 to 28.9	12
29 to 31.9	12
32 to 34.9	8
35 to 37.9	4
38 to 41	3

4.

Craft	*Apprentices*
Carpentry and joinery	80
Wood machining	24
Painting	48
Brickwork	34
Plumbing	54

5.

A	20%
B	40%
2nds	30%
R	10%

6. Sample of 100 trees. Range 1 m to 1.6 m girth. Two thirds 1.3 m or more girth.

7. London garage dealt with 40% of the work (700 cars), Brighton garage

approx. 31% (550 cars), Guildford approx. 22% (375) and Petworth approx. 7% (125).

8. Sample of 100 bricks. 218 mm length had greatest frequency. 73% were between 217 and 219 mm. Only 10% were outside a 4 mm range about 218 mm.

REVISION EXERCISE E

1.

Marks	*Frequency*
0 to 9	1
10 to 19	2
20 to 29	3
30 to 39	6
40 to 49	6
50 to 59	10
60 to 69	7
70 to 79	2
80 to 89	2
90 to 100	1

2.

Craft	*Pie chart*	*Bar chart*
Carpentry	144°	40%
Brickwork	100.8°	28%
Plumbing	57.6°	16%
Plastering	14.4°	4%
Painting	43.2°	12%

4.

Component	*Pie chart*
Maintenance	90°
Fuel	162°
Spares	72°
Insurance	36°

8.

Strength in N/mm²	*Frequency*
300 to 319	3
320 to 339	24
340 to 359	26
360 to 379	22
380 to 400	5

Exercise 15

1. $x = 140°, y = 40°, z = 140°$ 2. $x = 36°$
3. $x = 30°, y = 60°, z = 120°$ 4. $a = 80°, b = 100°, c = 80°$
5. $a = 110°, b = 110°, c = 70°, d = 70°, e = 110°, f = 110°$, $g = 70°$
6. $q = 40°, p = 140°$
7. $p = 95°, q = 85°, r = 95°, s = 95°, t = 95°, u = 85°$
8. $x = 140°$ 9. 53°, 133° 30′ and 81° 05′
10. ∠AOD acute, ∠AOB right, ∠AOC reflex, ∠BOD = 60°, ∠BOC = 120°
11. 3 12. 4

Exercise 16a

1. Obtuse, 30° 2. 45° each 3. c, 45° 5. 53° and 76°
6. 72° 30′, 6° 50′, 63° and 37° 15′
7. 96° 50′, 7° 30′, 30° and 154° 10′

Exercise 16b

1. a) 51.42 b) 65 c) 170 d) 8.746
2. 3.75 3. 100 mm 4. $1\frac{1}{12}$ m 5. 5.56 m

6. 12 m and 24 m 7. 3 m 8. 1.626 m 10. CFD

11. a) ∠A = 48°, ∠B = 54°, ∠C = 78°
b) ∠A = 75°, b = 36.5 mm, c = 45 mm
c) ∠A = 36°, ∠B = 44°, c = 96 mm
d) ∠B = 34°, ∠C = 56°, a = 65.5 mm
e) ∠B = 32°, ∠C = 58°, c = 59.5 mm

Exercise 17a

1. 250 mm 2. 4.32 m 3. 350 mm 4. 2.4 5. 60°
6. ∠PBD = 45°, ∠PDB = 30° 7. ∠ADC = 25°, ∠PAD = 105°
8. ∠BOC = 150°, ∠OBC = 15° 9. ∠BDC = 40°, ∠CBD = 35°
10. 132.3 mm 11. 2.4 m 12. 47.7 mm and 31.25 mm

Exercise 17b

1. 57° 18′ 2. a) 0.6545 rad b) 3.0543 rad c) 3.6757 rad
3. a) 50° 08′ b) 207° 59′ c) 327° 44′
4. a) $\pi/18$ b) $\pi/12$ c) $\pi/5$ d) $5\pi/9$ e) $5\pi/4$
5. a) 135° b) 12° c) 72° 6. 2.304 m
7. Sector; a) 1.25 rad b) 71° 37′ 8. 41° 40′
9. a) 0.098 rad b) 5° 36′ (one complete swing comprises two arcs)
10. 2.255 rad, 129° 11′

Exercise 18a

1. Parallelogram 2. Trapezium 3. Rhombus
4. Rhombus 5. Rectangle 6. 2.09 m
7. 50 mm 8. 1.5 m and 2.6 m, 1.95 m²
9. 287.1 m² 10. 992 m² 11. 26.82 m²
12. a) 4.909 m² b) 48.275 m² c) 17.9 m²

Exercise 18b

1. 5.89 m², 1.104 m³ 2. 54.66 cm² 3. 2.059 kg
4. 4.3 mm 5. 2 517 168 m³ 6. 17.35 m³
7. 4 922 m³ 8. 98.13 mm, 2.82 kg 9. 0.5236 ($\frac{1}{6}\pi$)
10. 0.5 11. 810 000 litres

REVISION EXERCISE F

1. 103.9 mm, 60°

2.

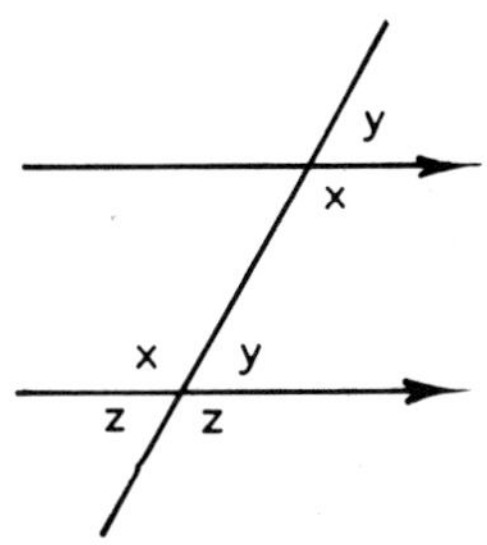

4. a) 74° 30′, 164° 30′
b) 62° 15′, 152° 15′
c) 8° 40′, 98° 40′
d) 45°, 135°
5. Obtuse angled, 30°, c
6. 135°, $22\frac{1}{2}$°, $22\frac{1}{2}$°
7. a) ∠B = 30°, c = 69.28 mm

b) $\angle C = 41°, a = 111.3$ mm

c) $\angle B = \angle C = 45°, a = 70.7$ mm

8. 109.6 mm 9. 21.54 m 10. 4.924 m 12. 318.3 mm

13. 523.6 mm 14. $\angle PSR = 50°, \angle XPS = 15°$ 15. 9.68 m

16. a) 0.921 rad b) 1.558 rad c) 2.793 rad

17. a) 85° 57′ b) 217° 43′ c) 329° 27′

18. a) $\pi/15$ b) $\pi/10$ c) $2\pi/3$

19. 1.125 rad (64° 27′) 20. 300.6 mm 21. 1.798 m

22. 6.713 m, 2.8 m^2 23. 491 mm^2, 89.27 mm 24. 0.583 m^2

25. 14.14 mm 26. 40 mm, 37.7 cm^3, 98 g 27. 405.4 ml

Exercise 19a

1. 35° 2. 37° 3. 20° 4. 73° 5. 51° 6. 17°

7. sin 20° = 0.342
   cos 20° = 0.940
   tan 20° = 0.364

8. sin 40° = 0.643
   cos 40° = 0.766
   tan 40° = 0.839

9. sin 50° = 0.766
   cos 50° = 0.643
   tan 50° = 1.192

10. sin 65° = 0.906
    cos 65° = 0.423
    tan 65° = 2.145

Exercise 19b

1. 0.4557 2. 0.5398 3. 0.9588 4. 0.7173

5. 0.1104 6. 0.1495 7. 0.9739 8. 0.1097

9. 3.1398 10. 68° 42′ 11. 87° 04′ 12. 45° 11′

13. 18° 40′ 14. 40° 14′ 15. 17° 06′ 16. 63° 38′

17. 26° 22′ 18. 60°

Exercise 19c

1. $\angle A = 35° 52'$, $\angle C = 54° 08'$, $b = 102.4$ mm, area = 2 490 mm^2

2. $\angle C = 64° 50'$, $a = 70.47$ mm, $b = 165.7$ mm, area = 5 285 mm^2

3. $\angle B = 39° 33'$, $a = 4.42$ mm, $c = 5.73$ mm, area = 8.07 mm^2

4. $\angle A = 42° 58'$, $\angle B = 47° 02'$, $b = 374.6$ mm, area = 65 367 mm^2

5. 9.02 m 6. 1 563 m^2 7. 97 mm 8. 45 m, 26 m

Exercise 20

2. cos 27° = sin 63°, cos 31° = sin 59°

REVISION EXERCISE G

The calculated ratios for questions 1 to 4 may be checked from mathematical tables.

1. $\angle A = 32°$, $\angle C = 58°$, $b = 65.19$ mm

2. $\angle C = 55°$, $a = 36$ mm, $c = 71$ mm

3. $\angle B = 32°$, $\angle C = 58°$, $a = 51$ mm

4. $\angle B = 19°$, $b = 30$ mm, $c = 85$ mm

5. 0.4617   6. 1.3335   7. 0.8587   8. 0.5978   9. 2.4963

10. 0.9630   11. 40°   12. 23° 30′   13. 51°

14. 51° 52′   15. 57° 07′   16. 44° 27′   17. 20°

18. ∠A = 33° 27′, ∠C = 56° 33′, b = 10.43 mm, area = 25.01 mm^2

19. ∠A = 42° 04′, ∠B = 47° 56′, b = 12.4 mm, area = 42.78 mm^2

20. 1 988 mm^2 21. Base = 50.1 mm, angles = 70° 30′, area = 1 770 mm^2

22. 78.78 m   23. Dia = 132.7 mm, chord = 74.2 mm

24.

0	$\pi/6$	$\pi/3$	$\pi/2$	$2\pi/3$	$5\pi/6$	π
1	0.866	0.5	0	−0.5	−0.866	−1

25. 38° (nearest degree)   26. 56° (nearest degree)

A checklist of books in the Newnes–Butterworths Technician Series

NEWNES–BUTTERWORTHS TEC TECHNICIAN SERIES

MATHEMATICS FOR TECHNICIANS 1

FRANK TABBERER, Chichester College of Technology

This is an introduction to mathematics for the student technician, intended especially to cover mathematics at level one in TEC courses (core unit U75/005). The presentation will create an interest in the subject particularly for those students who have previously found maths a stumbling block. There are frequent examples and exercises, with a summary and revision exercise at the end of each chapter.

CONTENTS: Manipulating numbers. Calculations. Algebra. Graphs and mappings. Statistics. Geometry. Trigonometry.

192 pages May 1978 0 408 00326 X

MATHEMATICS FOR TECHNICIANS 2

FRANK TABBERER, Chichester College of Technology

This will cover mathematics at level two in TEC courses (units U75/012 and either U75/038 or U75/039), for those who have completed (or gained exemption from) the work in *Mathematics for Technicians 1*. It will include the alternative schemes of work allowed in the second stage of level two. The presentation and systems of examples and exercises will be similar to those in the first volume.

September 1978 0 408 00371 5

PHYSICAL SCIENCE FOR TECHNICIANS 1

R. McMULLAN, Willesden College of Technology

This is intended for students studying the Physical Science level one unit of programmes leading to TEC certificates and diplomas. The text meets the requirements of the standard TEC syllabus for physical science (unit U75/004), a core unit of courses in building, civil engineering, electrical engineering and mechanical engineering. Attention has been paid to the visual presentation of the text, which is illustrated with diagrams and examples. Important concepts and formulae are clearly highlighted as an aid to learning and revision.

CONTENTS: Introduction. Fundamentals. Force and materials. Structure of matter. Work, energy, power. Heat. Waves. Electricity. Force and motion. Forces at rest. Pressure and fluids. Chemical reactions. Light. Index.

96 pages May 1978 0 408 00332 4

ELECTRICAL PRINCIPLES FOR TECHNICIANS 2

S. A. KNIGHT, Bedford College of Higher Education

Easy to read and in close conformity with the TEC syllabus, this book is intended primarily to cover TEC unit U75/019, Electrical Principles 2, an essential unit for both telecommunications and electronics students. The text includes examples, worked out for the reader, as well as problems for self-assessment, answers to which will be found at the end of the book. SI units are used exclusively throughout.

CONTENTS: Units and definitions. Series and parallel circuits. Electrical networks. Capacitors and capacitance. Capacitors in circuit. Magnetism and magnetisation. Electromagnetic induction. Alternating voltages and currents. Magnetic circuits. Reactance and impedance. Power and resonance. A.C. to D.C. conversion. Instruments and measurements. Alternating current measurements.

144 pages May 1978 0 408 00325 1

ELECTRONICS FOR TECHNICIANS 2

S. A. KNIGHT, Bedford College of Higher Education

Provides an introduction to the basic theory and application of semiconductors. It covers the essential syllabus and requirements of TEC unit U76/010, Electronics 2, though some additional notes have been added for clarity. The text includes examples and self-assessment problems.

CONTENTS: Thermionic and semiconductor theory. Semiconductor and thermionic diodes. Applications of semiconductor diodes. The bipolar transistor. The transistor as amplifier. Oscillators. The cathode ray tube. Logic circuits. Electronic gate elements.

168 pages June 1978 0 408 00324 3

BUILDING TECHNOLOGY 1 & 2

JACK BOWYER, Croydon College of Arts and Technology

These textbooks are primarily intended for the building technician taking TEC B2 construction courses. The clarity of text and illustrations should also, however, appeal to students of architecture and quantity surveying who need a good solid grounding in building construction.

BUILDING TECHNOLOGY 1

CONTENTS: The building industry. Site investigation, setting out and plant. Building elements: practice and materials. The substructure of building. The superstructure of building. Appendix: Building Standards (Scotland) Regulations 1971–75.

96 pages March 1978 0 408 00298 0

BUILDING TECHNOLOGY 2

CONTENTS: First fixing joinery and windows. Services and drainage. Finishes and finishings. Second fixing joinery and doors. Site works, roads and pavings. Appendix: Building Standards (Scotland) Regulations 1971–75.

96 pages May 1978 0 408 00299 9

HEATING AND HOT WATER SERVICES FOR TECHNICIANS

KEITH MOSS, City of Bath Technical College

By a system of nearly 200 worked examples, the author describes the routine design procedures for heating and hot water services in commercial and industrial buildings. Primarily intended for student HVAC technicians (TEC sector B3), it will also be useful for other students in sectors B2 and B3, and as a revision aid for experienced HVAC technicians encountering a change from Imperial to SI measurement.

CONTENTS: Heat energy transfer. Heat energy requirements of heated buildings. Heat energy losses from heated buildings. Space heating appliances. Heat energy emission. Heating and hot water service systems. The feed and expansion tank. Three-way control valves and boiler plant diagrams. Steam generation. Steam systems. Preliminary pipe sizing. Circuit balancing. Hydraulic resistance in pipes and fittings. Proportioning pipe emission. Hot and cold water supply. Circulating pumps. Steam and condense pipe sizing. Heat losses using environmental temperature. Medium and high pressure hot water heating. Index.

168 pages July 1978 0 408 00300 6

NOTES

NOTES

NOTES

NOTES